Are We Living in a Simulation or in a Material World Operated by a Simulation?

by
Alexander Popoff

Are We Living in a Simulation or in a Material World Operated by a Simulation?

Copyright © 2021 by Alexander Popoff
All Rights Reserved

Contents

1. Fermi Paradox: Where Are the Aliens?	5
2. The Fatima Miracle: Mass Mind Control	14
3. Near-Death Experiences Suggest Simulation	22
4. Material World Reflecting a Simulation	32
5. Reincarnation: Eternal Soul or Digital Technology	38
6. Mind Beyond the Brain	46
7. Man of Future	52
8. The Simulation Guides the Animals, Too	61
9. The Omnipotent Simulation Plays Games	67
10. Creativity, Intuition, and The Genius	90
11. Psychokinesis Parties	117
12. Creating Universes and Educational Environment	122
13. Precognition of the Future and Determinism	149
14. Adolf Fritz: The Surgeon from The Beyond	183
15. Psychic and UFO Healings	191
16. Dinosaurs on the Moon or Engineered Extinction	206
17. Dowsing	232
18. Transformation: Entering a New World or a Hell	237
19. Various Cases	248
20. UFOs: Mind Control, Materializations, or Alien Spacecraft?	260
Afterword: designer civilization in a designer world	314

"…each man explains in his own way the fact that the human will is not free…Everything is determined…by forces over which we have no control…for the insect as well as for the star. Human beings, vegetables, or cosmic dust, we all dance…intoned in the distance by an invisible player."

Albert Einstein

In search of the invisible player

Chapter 1
Fermi Paradox: Where Are the Aliens?

The Universe is homogeneous (it looks the same everywhere on a large scale) and very old. There is every reason to believe that it should be teeming with legions of space civilizations. Numerous solutions have been proposed to the key question of why there is no hard evidence of the existence of any extraterrestrial intelligence. This great number of proposed hypotheses, which flood the pages of books, magazines, academic journals, newspapers, and the internet, indicates that there is still no satisfactory explanation of this fundamental conundrum.

However, if we assume that for some reason the space civilizations in our Universe started at about the same time (the equal start hypothesis), we get a very elegant resolution to the Fermi paradox: the Universe is teeming with sentient beings, but since they are at almost the same level of development, most of the numerous intelligent races still have not contacted or have not found evidence of the existence of other advanced creatures. They, just like humans, are making their first steps into deep space. The leading intelligences already travel across the neighborhood of their star system, but they face a lot of problems: financial, biological, technological, and so on, which limit their space expansion activities.

What could be the reason for such an equal start to intelligence?

The equal start hypothesis presupposes a factor or factors providing such an equal start. It could be something that "forbids" the emergence of sophisticated life before a certain point in time, like frequent gamma-ray explosions, constant impacts from large comets and asteroids, massive volcanic eruptions, disastrous climate changes, and other devastating natural events on a universal and galactic scale. Even if primitive life becomes widespread, for billions of years it could not make the step to higher life forms and intelligence.

However, there is also a more complex explanation. There are clues and evidence that life in our Universe is orchestrated by a superior intelligence.

But what would be the point of the almost simultaneous emergence of the sentient space races? What could be the plans of a superior godlike intelligence able to control the living creatures in the entire cosmos?

All life in our Universe is based on hard competition in order for Nature to accelerate evolution, thus providing (bio)diversity, quality, and quantities of advanced civilizations.

The intelligent races can compete, cooperate, and progress successfully only if they are at about the same level of development. The later-emerging space civilizations could hardly survive a real competition with sentient beings, who would be far ahead of them.

The level of competition should be defined very well.

If the competition is too tough, the rate of death and destruction resulting from rivalry (wars, revolutions, religious clashes, crime, all sorts of accidents, jealousy, and so on) becomes too high and the development will eventually slow down: industrial, financial, infrastructural, life's losses, and other losses turn into stopping factors. This is especially important when intelligent species develop immensely destructive technologies like nuclear, chemical, biological, Artificial Intelligence, robotic, nanotechnological, and other weapons of mass destruction—even if the intelligent species survive, the devastation and havoc could be so huge that such civilizations might not recover in a reasonable period of time and they would be assimilated or destroyed by their competitors—human or alien.

If the level of competition is too easy, then the destruction and the frequency of death within a given population as a consequence of competition is significantly reduced and people enjoy a comfortable life; however, the rate of development would be insufficient, and such civilizations might not survive the rivalry.

Biological life and all human history on Earth are actually a story of tough competition and the struggle for survival: between the Cro-Magnons and the Neanderthals, between individuals, tribes, states, armies, companies, species, religions, languages, etc.

Hundreds of millions of people have died as a result of the competition on Earth.

According to *The New York Times*, July 6, 2003, "Estimates for the total number killed in wars throughout all of human history range from 150 million to 1 billion."

Billions more will die in the years to come.

Nothing and no one can escape competition.

The refusal to compete and evolve means certain death to any civilization.

If future humans, in whatever physical form they are, develop below expectations, they will be deleted by our space competitors. Only the best will survive, the rest will be annihilated. When it comes to fast evolution, there is no mercy.

The civilizations can compete successfully only if they are at about the same level of development.

Could the prominent Exercitus Romanorum (the army of the great Roman Empire) defeat a modern army? No way!

Roman scientists couldn't even imagine the science, technologies, and weapons of our present armies. They had absolutely no idea about nuclear bombs, which could destroy all creatures on a given planet. Roman military specialists did not have the slightest notion about computers, radios, radars, satellite communications, observation satellites, common army rifles, pistols, machine guns, modern artillery, tanks, helicopters, airplanes, missiles, nuclear, and regular submarines that can stay underwater for months and cross oceans… The list of all available modern weapons is very long.

Analogously, we cannot even imagine the science, technologies, weapons, and the power of an alien civilization only 2,000 years ahead of us, especially taking into account the exponential development of modern science and technology.

But what if the technological difference is not 2,000 years, but much smaller—say, only 27 years. It is a minor difference.

World War I ended in 1918. World War II ended in 1945. The technological time difference is 27 years.

Could the armies of World War I overcome the armed forces of World War II? No way!

If you compare the two armies, separated by a period of only 27 years, you will see that a World War I army cannot defeat a World War II army. All weapons of World War II armies were much better than that of World War I armies. There were also new weapons like radar, anti-aircraft rockets, bazookas, bomber aircraft, anti-tank guns, cipher devices (Enigma machine), rockets (V-2 and Katyusha, the multiple rocket launcher), anti-aircraft cannons, automatic rifles, bombsights that allowed high-altitude precision bombings above the range of the anti-aircraft cannons, sophisticated radio communication, airplanes armed with cannons, etc. The World War II army possessed the ultimate weapon of the time—the atomic bomb, an apotheosis of science, technology, destruction, and terror, and a constant threat to human extinction.

Researchers normally suggest great age differences between the cosmic races—millions, hundreds of millions, or even billions of years, taking into account that the Universe originated about 13.8 billion years ago and assuming that sentient creatures are emerging at a constant rate.

The late-emerging civilization cannot survive competition with superior space races.

Stephen Hawking stated that, "Life on Earth is at the ever-increasing risk of being wiped out by a disaster such as sudden global warming, nuclear war, a genetically engineered virus or other dangers… I think the human race has no future if it doesn't go into space. There are too many accidents that can befall life on a single planet."

All space civilizations will follow such survival strategy and will colonize space, fighting with each other for resources, Goldilocks planets, and convenient natural satellites. If humanity wants to survive, it also should compete (even wage grand-scale and local wars) for resources and space bodies to be colonized.

The equal start hypothesis suggests that accurate timing is very important for complex systems, because instead of a Universe filled to the brim with fit, intelligent races, there might be one or only a few of them, which would never reach the high level of development as a legion of hard-competing civilizations would.

If the emergence and the development of space civilizations is orchestrated, it is reasonable to expect that the Universe, life, and intelligence are subject to a creative external guiding agency, plus chance events, and evolution through natural selection. Possibly the plan of the external agency is to create a great many space civilizations developing as fast as possible.

According to this scenario, evolution is both natural and assisted. Humanity is a designer civilization.

Let's start looking for evidence that we are really under the control of a superior intelligence.

Chapter 2
The Fatima Miracle: Mass Mind Control

The Fatima phenomenon is among the best documented cases, confirming mind control over the humans by an external agency.

In 1917, some very striking events happened near the small town of Fatima in Portugal. A large crowd of 100,000 people observed the Miracle of the Sun. The events were officially accepted as a miracle by the Roman Catholic Church. Some researchers claim that it was a mass UFO sighting or a rare meteorological event.

On October 13, 1917, the lawyer Dr. José Garrett, armed with binoculars, was standing on the rim of a hollow field called Cova da Iria with a commanding view of the entire area. He counted more than a hundred thousand people there.

Dr. Garrett wrote a detailed report on this event. Newspaper reporters were also in attendance. They wrote numerous articles and took testimony from many people who witnessed extraordinary solar activity and other strange events.

The large crowd had gathered there because several times, three young shepherd children had seen a female apparition (some people believed it was the Virgin Mary) who predicted a few months earlier that at high noon on that day she would perform a great miracle "so that all people may believe!"

The big day was gloomy and rainy. The sky was covered with clouds and it was still raining.

The crowds were making a tremendous noise that could be heard far away.

At noon the rain stopped and the Sun started breaking through. It began to wander within the circle of the receded clouds. It seemed pale, like a ball of snow. Then the solar disk began to rotate about its center at a rather high velocity.

The most astonishing thing was that the people were able to stare at the solar disc for a long time, brilliant with light and heat, without hurting their eyes or damaging their retinas.

And suddenly everything around turned into different colors.

The Sun was spinning rapidly like a fire-wheel, throwing beams of many colors. The people and everything around were stained with the colors of the rainbow. The Sun seemed literally to dance in the sky.

A witness said, "The seconds seemed like hours, so vivid were they."

The people were enchanted by a remarkable spectacle in the sky of a kind they had never seen before.

Then the Sun turned blood red; it suddenly stopped spinning and seemed to come down in a zigzag, strongly radiating heat, threatening to burn and smash the Earth. It came rapidly closer and closer, with increasing size and brilliancy. A terrible moment!

The people cried, "It comes down! It comes down!"

Most of the panicked people thought their last hour had come. They were weeping and expecting the end of the world at any moment. The crowd felt the terrible heat of the huge, fiery Sun that was coming down.

The people believed it was the end of the world.

Some were losing their senses.

The crowds were making a tremendous noise. Most of them were crying out to God for mercy. Many were confessing their sins aloud, asking God to pardon their sins. Parents were throwing themselves protectively over their children. A few were trying to run away from the falling Sun. Some were simply paralyzed with fear.

The great ball of fire stopped just as it was about to crash upon the Earth, and the Sun was back in its usual place.

The great terror of being burned and smashed by the huge, fiery Sun was followed by an enormous feeling of joy and relief.

Many people were on their knees, crying, praying really hard, or calling on the Virgin Mary. They were fully convinced—it was a miracle, a real miracle! There is a God in the Heavens! Atheists were converted! People shouted, "I believe! I believe! I believe!"

One witness later said, "My mother who had a large tumor in one of her eyes for many years, was cured. The doctors who had attended her said they could not explain such a cure."

Despite the attempts of photojournalists, the phenomenon itself could not be photographed. In the pictures we can see only crowds looking at the sky.

The solar phenomenon was not observed in any observatory.

The religious people believe it was a miracle, the UFO enthusiasts claim that this was a huge alien spaceship, some researchers suggest that this was some strange meteorological phenomenon.

It wasn't an extraterrestrial spacecraft because some of the witnesses observed only cloudy skies, and no UFOs or some phenomena at all. Second, in the pictures of the photojournalists present we see again only cloudy skies.

It wasn't some meteorological phenomenon. How was it possible for small illiterate shepherds to predict several months upfront the appearance of a unique meteorological phenomenon at noon on the 13 of October near Fatima? Second, why not all witnesses observed the meteorological phenomenon but only cloudy skies? Why couldn't the photojournalists take pictures of the unique meteorological phenomenon?

The Miracle of the Sun at Fatima, many religious miracles, numerous UFO sightings, and some paranormal phenomena, while witnessed by most of the people present, are usually never witnessed by all of them. Some of the Fatima witnesses that day, including believers, reported that they saw only cloudy skies.

These "anti-witnesses" are actually very useful for researchers because we know that such striking events didn't really happen but are controlled hallucinations (visual and emotional) created by a superior entity.

Hallucinations could also originate in the unconscious mind of individuals as a product of their vivid imagination or pathology—but a group of such people never have the same hallucinations simultaneously.

The Fatima phenomenon was obviously a carefully planned and deliberately executed demonstration of mass control over the minds of people by an external agency.

The researcher Andrija Puharich wrote in his book about Uri Geller that once when driving in the desert, Puharich, Geller, and another witness saw a huge spaceship but the three military personnel in the front seat were unable to see it. Andrija Puharich also recorded other occasions when he and Uri Geller could see UFOs which the other individuals present could not observe.

The next case is from the book "Extrasensory Ecology: Parapsychology and Anthropology," by Joseph K. Long, published in 1977.

"In the early 1950s, Mr Bharati heard that there was a fakir in the Almore district of the Himalayas who was willing to levitate for anyone who was curious to watch. About fifty people showed up for the performance, including Mr Bharati...By six the following morning the crowd was well satisfied with what they had seen, but Mr Bharati had seen nothing. When he spoke with some of the other spectators, he learned to his astonishment that they had seen the fakir rise at least nine feet into the air."

When a group of people witnesses a paranormal event, including UFOs, sometimes they observe different objects and a different chain of events. In the following case the first witness saw a flying saucer, the second one just a regular bus. The story is from the book *Vedic Cosmography and Astronomy* by Richard L. Thompson. In 1971, the Brazilian Paulo Gaetano was driving with his friend Elvio when he saw a flying saucer following them. Gaetano told Elvio what he saw but Elvio could see only a bus following them. According to Gaetano, the car engine stopped and a red beam of light was projected at the car that caused the door to open. Several small extraterrestrials appeared. They abducted Gaetano and examined him medically in the alien spacecraft. They lowered from the ceiling an apparatus that looked like an X-ray machine.

Elvio saw nothing of the flying saucer but only a bus which was keeping at reasonable distance behind the car. He stated that the car slowed down and stopped. Paolo opened the door, went out, and fell to the ground behind the car. Elvio managed to get Paolo on his feet and both went by bus to the nearby town. Elvio didn't see the flying saucer, the red beam, or extraterrestrials abducting Gaetano.

Many such events happened only in the controlled minds of the witnesses.

In some cases when the experiencers tried to touch an alleged non-terrestrial spacecraft, alien robot, or extraterrestrials, or embrace the Virgin Mary, they couldn't feel anything—their hand passed right through them as if they were a ghost image or phantasm. In other cases the contactees touched the alien spaceship and felt the cold or warm metal surface, but when your mind is under control, you can't be sure of anything.

Chapter 3
Near-Death Experiences Suggest Simulation

Maybe the Near-Death Experience (NDE) phenomena could shed some additional light on the mind control riddle.

A near-death experience, numerous instances of which have been widely reported (many of them by physicians), is an experience most often associated with impending death or a severe life-threatening illness or event, encompassing multiple possible sensations such as detachment from the body, feelings of levitation, total serenity, security, and warmth, a life review, the presence of a bright light and creatures of light, meeting dead relatives and friends, etc.

Many near-death patients reported being able to watch and recall events during the time of cardiac arrest and resuscitation procedures. They were clinically dead with flatlined brain activity, but later these people can describe with accuracy what the doctors and the nurses have done on their bodies in the hospital room, watching them and their bodies from above. Their claims were later confirmed by hospital personnel.

Raymond Moody, a philosopher, psychologist, and physician wrote in his book *Life After Life*, "Several doctors have told me, for example, that they are utterly baffled about how patients with no medical knowledge could describe in such detail and so correctly the procedure used in resuscitation attempts, even though these events took place while the doctors knew the patients involved to be 'dead.'"

According to Pim van Lommel, cardiologist and researcher, some patients remembered details of their conditions and events during their cardiac arrest despite being clinically dead with flatlined brain activity.

Raymond Moody and Paul Perry reported the following cases in their book *The Light Beyond*.

The authors wrote that they have examples of people who had out-of-body experiences during their resuscitations and were able to leave the operating room and observed relatives in other parts of the hospital.

"Another woman had an out-of-body experience and left the room where her body was being resuscitated. From across the hospital lobby, she watched her brother-in-law as some business associate approached him and asked what he was doing in the hospital.

"Well, I was going out of town on a business trip," said the brother-in-law. "But it looks like June is going to kick the bucket, so I better stay around and be a pallbearer."

A few days later when she was recovering, the brother-in-law came to visit. She told him that she was in the room as he spoke to his friend, and erased any doubt by saying, "Next time I die, you go off on your business trip because I'll be just fine."

He turned so pale that she thought he was about to have a near-death experience himself.

How could clinically dead patients with flatlined brain activity receive, process, and memorize visual and auditory information about the cardiac arrest and the resuscitation procedures? How could they see and hear what was happening in the waiting room?

A probable explanation is that part of what we consider human consciousness is not in the brain. And second, there is another observational system that processes and records visual and auditory information which is delivered to the patients when they are conscious again. The normal visual vantage point of the humans is through their physical eyes. The usual vantage point during NDEs is above the human body, close to the ceiling; it could also move freely through closed doors, walls, bodies of people, etc.

What are these mysterious "eyes" that "float" above the patients that can move through walls, closed doors, and people? And how can the "dead" patient see events and hear conversations in other rooms of the hospital?

In a computer-generated simulation one can choose freely the perspective from which a character or a game player looks at the artificial environment. The displayed vantage point of the viewer allows the digitally simulated space to be seen from various angles and observation points, as if looking through a mobile surrogate set of eyes or a virtual camera. There is no need for material eyes and cameras. One can also move the virtual camera at will through closed doors, walls, and bodies of people.

Our reality seems to be material but it also looks like a simulation controlled by an extremely complex (far beyond our comprehension) super-intelligent, sophisticated data processing System inextricably coupled with the Universe, living creatures, and the intelligences. The System is rendering a simulation and our world is driven by the simulation.

The System is still beyond the reaches of our scientific instruments.

The System is not a substitute for God. It was created by a technologically advanced superior intelligence. The System is not a computer. Modern supercomputers compared to the System are very primitive devices.

The new idea suggests that we live in a material world reflecting a simulation controlled by a data processing System, natural laws, and most importantly, by the will and philosophy of the superior intelligence operating the Universe.

We are living in a material world reflecting a sophisticated simulation. The Grand Hermetic Principle of Correspondence says, "As above, so below." Finally, the meaning of the famous occult axiom is revealed: above is the guiding simulation created and controlled by a superior intelligence, below is the "subjugated" material world with all stars, planets, humans, and aliens.

The System and the simulation are technological products. Even though humans look like a result of long biological evolution and chance events, they are a biotechnological product. Our lives, destinies, and future are controlled products of the technological System.

Even our Universe is a technological product created by a superior intelligence.

The NDEs show that part of the human mind is not in the brain but in the System.

In 1998 Kenneth Ring and Sharon Cooper published an article in the "Journal of Near-Death Studies" about blind people (including those blind from birth) who have vividly visual NDEs or out-of-body experiences not associated with NDEs. There is no other explanation of such cases except that there is a System controlling the environment and minds of humans. It is observing the surroundings and sometimes delivers visuals into the mind of blind individuals.

In her book *On Children and Death* Elisabeth Kübler-Ross wrote about blind children who can see during NDEs.

"We, naturally, checked these facts out by testing patients who had been blind with no light perception for years. To our amazement, they were able to describe the color and design of clothing and jewelry the people present wore. I am sure no scientist could call this a projection. When asked how they could see, people described it with similar words: 'It is like you see when you dream and you have your eyes closed.'"

During the life review near-death experiencers see real events from their lives as though they are scenes from a movie about themselves. They view themselves from a third-person perspective, the perspective of the System.

A NDE witness remembers, "It wasn't like I was watching it all from my perspective at the time. It was like the little girl I saw was somebody else, in a movie, one little girl among all the other children out there playing on the playground."

Most near-death experiencers don't see their life reviews from their own perspectives, but from the perspective of a virtual camera of the controlling System.

Robert Sullivan, NDE researcher from Pennsylvania who specializes in NDEs by soldiers during combat, interviewed a World War II veteran who experienced three-hundred-sixty-degree vision while running away from a German machine-gun nest. Not only could he see ahead as he ran, but he could see the gunners trying to take aim on him from behind.

In moments of danger some individuals could see themselves from above or from multiple perspectives simultaneously.

There are war cases when during very intense combats some participants could see the battle-field from above, the time slowed down, and the soldier could see every flying bullet, every shell, every explosion in slow motion. That way they survived the deadly battle chaos.

There are many other examples when people could see from the perspective of the System which can also manipulate our perspective of time: "Time seemed to slow."

The life review during NDEs proves that the entire life of people is observed by the System and is recorded. It can be seen not only by near-death experiencers but also by some psychics.

Michael Talbot wrote about just such highly a sensitive female in his book *The Holographic Universe*. She often sees what looks like a little transparent movie going on around a client's head.

Some researchers and ordinary people (religious and nonreligious) suggest that NDEs are evidence of consciousness (eternal soul) existing separately from the body, which could be explained as the existence of life after death.

NDEs are not a proof of afterlife. The incredibly vivid, real, and mind-blowing experiences are just records of the System controlling our minds, including our emotions. Probably this mind manipulation has to convince humans to believe in the afterlife and not to fear death, or at least not so much. And that their insignificant, trivial, and often failed lives will have a real value in the reincarnations to come.

This NDE defense and manipulation mechanism is not only for dying people but also for the living people. The survivors tell the living ones their wonder stories about rebirths, contact with deceased relatives and religious super-beings, eternal life, immortality of the soul, miracle levitation, and how wonderful it is to live in the splendid world of the dead people. These stories become sources for articles, books, movies, social communication, etc., and they turn into a system of belief in the immortality of the human soul, existence of gods and angels, saints, devils, etc.

Many survivors report that their NDEs are like dreams, difficult to put into words. Actually they are a type of dreams fabricated by the System, just like many cases of alien abductions, UFO encounters, religious miracles, seeing mythological creatures, saints, gods, angels, etc. These experiences are very emotional, impressive, and hard to forget.

Some NDE survivors have had horrifying experiences involving tortures by elves, giants, demons, Satan, etc. This is also part of the control mechanism: the good people will meet their relatives, friends, and angels. The bad ones are destined to meet demons and Satan. They will go into hell and will be tortured. The System provides evidence for existence of paradise with angels and good people, but also for hell with the devil and his horrifying servants. There is no paradise without hell, no God without the devil, no good without evil.

The System is the ultimate propaganda master.

Everything is possible in a material Universe reflecting a simulation: seeing dinosaurs and all sorts of monsters; levitation; teleportation of objects and people; poltergeist phenomena; mothmen flying without flapping their wings; materializations, dematerializations, shapeshifting, merging or splitting of UFOs; huge airplanes with almost no wings flying noiselessly; walking through walls; unexplainable healings; turning water into wine or lead into gold, etc., etc.

Chapter 4
Material World Reflecting a Simulation

"The Universe can be regarded as a giant quantum computer," writes professor Seth Lloyd of the Massachusetts Institute of Technology in his book *Programming the Universe.*

Lloyd suggests that the Universe itself is one big quantum computer producing what we see around us, and ourselves, as it runs a cosmic program.

"The universe is made of bits. Every molecule, atom, and elementary particle registers bits of information. Every interaction between those pieces of the universe processes that information by altering those bits...The universe is a quantum computer. This begs the question: What does the universe compute? It computes itself. The universe computes its own behavior. As soon as the universe began, it began computing."

Lloyd wrote that at first, the Universe produced simple elementary particles and established the fundamental laws of physics. Later it created galaxies, stars, planets, life, intelligent beings, and society. According to

Lloyd, the computational capability of our Universe explains one of the great mysteries of nature: how complex systems such as living creatures can arise from fundamentally simple physical laws.

"The computational universe necessarily generates complexity. Life, sex, the brain, and human civilization did not come about by mere accident."

According to the Wikipedia definition, digital physics is the idea that our Universe can be considered a vast digital computation device, or as the output of a deterministic or probabilistic computer program. The hypothesis that the Universe is a digital computer was proposed by German scientist Konrad Zuse in his book *Rechnender Raum* (translated into English as *Calculating Space*) in 1969. His greatest achievement was the world's first programmable computer which became operational in May 1941. Zuse proposed that the Universe is being computed by some sort of cellular automaton or other discrete computing machinery, challenging the long-held view that some physical laws are continuous by nature.

The term digital physics was employed by Edward Fredkin in 1978, who later came to prefer the term digital philosophy. Digital physics suggests that there exists, at least in principle, a program for a universal computer that computes the evolution of the Universe.

According to Fredkin's theory of digital physics, information is more fundamental than matter and energy. He believes that atoms, electrons, and quarks consist ultimately of bits; the behavior of those bits, and thus of the entire Universe, is governed by a single programming rule, "the cause and prime mover of everything."

Fredkin's digital philosophy contains several fundamental ideas: 1. Everything in physics and physical reality must have a digital informational representation. 2. All changes in physical nature are consequences of digital informational processes. 3. Nature is finite and digital.

According to pancomputationalism, everything is a computing system. It provides an understandable theory of everything. It suggests that the world is a computer or it can be described as a computer.

Pancomputationalists believe that biology reduces to chemistry which reduces to physics which reduces to the computation of information.

The new model of our Universe suggests that it is material, has characteristics of a giant quantum computer, and is driven by a simulation, which includes natural laws, and is operated by a data processing System created and controlled by a superior intelligence.

Science riddles like fine-tuning of the Universe, the UFOs and alien abductions, paranormal phenomena, and so on are possible not only in a computer-simulated world, but also in a material world reflecting a sophisticated simulation.

For some researchers the Universe is intelligent, conscious, or the Universe itself is some kind of gigantic intelligent organism. But it is not. It is just matter and energy. The System is intelligent. What is the System, we still don't know. The best approximation we can make is that it is a superior quantum data processing System running a simulation of the Universe but it is probably much more; it is far beyond our current knowledge and science fantasy.

What is the superior intelligence that created and rules the Universe? It is so advanced that we can't comprehend it. The numerous human gods are creations of the System and the superior intelligence. Because we can't comprehend the superior intelligence, it created the biblical God and the son of God on Earth—Jesus, an entity we can understand and which in many ways is like us. Jesus lived and suffered like all humans do. He is near to people's hearts. The supernatural entities God, Jesus, Buddha, etc. have their real power on Earth endowed to them by the System and the superior intelligence.

Why would someone start the simulation and the Universe from the very Big Bang (whatever it is) and why should the simulation and the Universe go through all phases of the evolving Universe? It takes a lot of time (billions of years) and is actually repeating something they already know and have gone through? Or maybe the simulation and our world really started 300 years ago. The Bible says the world is between 6,000 and 10,000 years of age. The previous history is fake, the fossils of prehistoric animals and the archeology are fake. They were implanted for us to study them. We don't know the plans and the philosophy of the superior intelligence that created the Universe, the life, and the sentient creatures. We still don't know when our Universe started—300 years ago or 13.8 billion years ago. In a material world reflecting a simulation it is difficult to determine when it started.

On the other hand, we should not forget that it is also possible that the Universe outside the Solar System is only a digital simulation.

This can explain the Fermi paradox of why we have no evidence of existing of extraterrestrial civilizations. Humans are alone. But I prefer to believe that the entire Universe is material with a great number of civilizations.

Chapter 5
Reincarnation: Eternal Soul or Digital Technology

Some mediums enter a deep trance, during which they appear to be taken over entirely by a new personality with a different name, pattern of speech, voice, gestures, etc.

Some researchers claim that mediums, shamans, psychics, healers, etc. are channeling spirits of dead people or other entities in a state of altered consciousness. Well, not really, only sometimes. In most cases their mind is in normal mode but they are different compared to ordinary people. They receive information from the System in the form of visions, thoughts, voices, feelings.

Some psychics and mediums fake entering into a trance in order to impress the sitters.

Many people are convinced that during spiritism séances they are communicating with their beloved dead relatives or friends because the medium talking with the dead individuals is reporting particulars not known to external parties; the channeled deceased person uses specific family jokes, words, phrases, and manner of speaking exactly corresponding to the person when he/she was alive. In many cases, the medium even begins to talk in a voice that sounds exactly like the dead individual.

The dead husband tells his bewildered widow through the medium where exactly he put his expensive gold watch the whole family is looking for. But dead people are not alive in some sort of eternal paradise. These are only digital records of their past lives revived by the System to convince baffled relatives and friends that their deceased loved ones are still alive, living somewhere else.

Not only psychics, mediums, contactees, and that sort of people receive channeled information, but also researchers, scientists, ordinary people; actually, all people are connected with the System. In the Fatima case a large group of about 100,000 people simultaneously received pictures, thoughts, and emotions of a miracle.

Reincarnation is the concept that the human soul or spirit, after biological death, can begin a new life in a new body. There is a huge amount of fiction and nonfiction advocating this doctrine.

According to William Walker Atkinson, author of the book *Reincarnation and the Law of*

Karma, published in 1908, "This fundamental belief may be expressed as the doctrine that there is in man an immaterial Something (called the soul, spirit, inner self, or many other names) which does not perish at the death or disintegration of the body, but which persists as an entity, and after a shorter or longer interval of rest reincarnates, or is reborn, into a new body—that of a unborn infant—from whence it proceeds to live a new life in the body, more or less unconscious of its past existences, but containing within itself the 'essence' or results of its past lives, which experiences go to make up its new 'character,' or 'personality.'"

The psychiatrist Ian Stevenson became known for his research into cases he considered suggestive of reincarnation. Over a period of forty years in fieldwork, he investigated three thousand cases of children who claimed to remember past lives.

Some individuals, believing that they reincarnated, provide descriptions of a life of a past person, and sometimes the accuracy of their information can be confirmed, fully or partially. In some cases the reincarnated people have considerable knowledge of places, houses, clothing, events, wars, arms, customs, traditions, etc. connected with the lives of their supposedly previous personality that they could not have learned normally. They sometimes inherit skills from their previous lives like handling a boat, writing short stories, novels, or poetry, climbing high in trees, drawing, singing, drumming, playing a guitar, etc.

Some cases include scars or birthmarks on the bodies of the previous person and the reincarnated.

The reincarnated could have preferences for specific foods and drinks characteristic to the past person. They also could have specific postures, gestures, or show unusual behavior such as writing and reading from right to left characteristic of the person from their alleged previous lives.

Many reincarnated have fears and phobias connected with their previous lives: fear of water because in her previous life she was a beautiful redheaded dancer and singer, and was drowned by a jealous lover; fear of Germans because he was killed in a concentration camp; fear of falling because she broke her leg badly in a rabbit hole running downhill as a kid and died two months later; fear of vehicles because he died in a car accident; fear of fire, airplanes, etc.

Most often the memories of a previous life fade over time. By eight to ten years of age, only some children still have such memories. In most cases they stop speaking about their previous lives when they turn seven to eight years old.

Some children reject their given names and claim the name of the previous personality.

Sometimes children reject their mothers and want their "real" mother to take care of them, actually the mother of the previous personality.

In some cases when a child sees a photograph of the personality from a previous life, he/she claims that this is his/her photography.

Kids' statements about their previous lives are usually stimulated by belongings, photographs, or persons related to the previous persons.

Girls consider themselves boys, and boys consider themselves girls, temporarily taking the gender of the previous personality.

The reincarnation cases suggest there is a system of inheriting personalities, something like a personality software. Reincarnation is the religious interpretation of the observation of many such cases of inherited personalities. The definitive scientific explanation is still lacking. The System is in charge of inheriting suitable personalities in order to accelerate the individual growth of people and evolution of mankind. That way the kids grow up intellectually and socially much faster. They can also receive specific skills from a previous person.

The new individual inherits some sort of "essence" (actually data) from the previous life of some person. Probably there are also synthetic personalities made of the lives and characteristics of two or three previous persons who were very successful. The System and humanity need political, social, intellectual, military, etc. leaders.

There are cases that confirm that this is inheritance of personalities and not reincarnation in the religious sense of the word. According to the reincarnation doctrine, the soul transmigrates at the moment of conception, the soul of the child (still a fetus) is an active participant in forming the body that it is to inhabit to fulfill its mission in life. The soul of the person to be "reborn" should enter into a new body—that of an unborn infant— and that should happen about nine months before the birth of the child, at the moment of conception. But there are cases, in which the reincarnated was born two months, six months, or even a week after the death of the previous personality.

That would mean the future child did not have a soul for many months, which is impossible, according to religious doctrines. The soul should enter into its new body at the moment of conception.

Most curiously there is a case when the reincarnated subject was born about five weeks before the previous personality died; this is the case of Ruprecht Schulz from the book *European Cases of the Reincarnation Type* by Ian Stevenson. One soul was in two bodies? Or two personality records in the System.

What makes us humans? What makes the huge difference between the humans and the chimps, taking into account the fact that both species are almost genetically identical—humans share about 99% of our DNA with chimpanzees? The System makes the difference! The bodies of the species are a result not only of the DNA but also of an inheritable invisible morphogenetic structure forming our organs and body (still undiscovered by the contemporary science) and personality "software" forming our mind and personality.

Science is indebted to interpretation of the personality inheritance observations that allow the System, some religions, cults, and

New Age lore to advocate the reincarnation doctrine.

Chapter 6
Mind Beyond the Brain

The idea that part of the mind is not in the brain but in the System is supported by some cases of psychiatry and neurology.

Michael Nahm, David Rousseau, and Bruce Greyson in their article "Discrepancy Between Cerebral Structure and Cognitive Functioning," published in 2017 in the *Journal of Nervous & Mental Disease* wrote:

"Neuroscientists usually suppose that human mental functions are created by the brain."

"Nevertheless, several cases involving brain dysplasias (abnormal cell development) and brain lesions (cell damage) indicate that large amounts of brain mass and its organic structures, even entire hemispheres, can be drastically altered, damaged, or even absent without causing a substantial impairment of the mental capacities of the affected persons."

Hydrocephalus is a condition in which an accumulation of cerebrospinal fluid occurs within the brain. The term hydrocephalus is derived from the Greek words "hydro" meaning water and "cephalus" meaning the head.

The authors presented several cases of persons with badly damaged brains but substantially unimpaired mental capacities.

A highly intelligent student of mathematics had a global IQ of 130 and a verbal IQ of 140 at the age of 25 but had "virtually no brain."

This student belonged to the group of patients classified as having "extreme hydrocephalus," meaning that more than 90% of their cranium appeared to be filled with cerebrospinal fluid.

Another interesting case is that of a 44-year-old woman with very gross hydrocephalus who had a global IQ of 98, yet worked as an administrator for a government agency and spoke seven languages.

In Leipzig, Germany, staff members of the Max Planck Institute for Human Cognitive and Brain Sciences recorded a similar case. A man was examined because of his headache, and to his physicians' surprise, he had an "incredibly large" hydrocephalus. Villinger, the director of the Cognitive Neurology Department, stated that this man had "almost no brain," only "a very thin layer of cortical tissue." This man led an unremarkable life, and his hydrocephalus was only discovered by chance.

It is astonishing that many patients can lead an ordinary life after a drastic procedure like surgically removing one hemisphere of the brain. They actually live with half brain.

The savant Kim Peek (1951–2009) had perfect memory: he forgot nothing he ever read and remembered complete melodies, even if he heard them only once. Most remarkably, his brain showed considerable malformations that included a deformed cerebellum (the part of the brain at the back of the skull), abnormalities of the left hemisphere, and the complete lack of the corpus callosum (the bundle of nerves that connects the two hemispheres). In addition, much of the skull interior comprised empty areas that were filled with cerebrospinal fluid, as in hydrocephalic subjects. Nevertheless, he memorized more than 12,000 books, apparently verbatim, the contents of which amounted to an encyclopedic knowledge in multiple areas of interest. Typically, he would read a page in eight to ten seconds and then turn to the next page. He even read two pages of smaller books such as paperbacks simultaneously, using one eye each for each page. Kim could read a page turned sideways or upside down.

Moreover, he had impressive calendar calculating abilities. Peek enjoyed approaching strangers and showing them his talent for calendar calculations by telling them on which day of the week they were born and what news items were on the front page of major newspapers that day.

His IQ was 87. The average IQ is 100.

He had encyclopedic knowledge in multiple areas of interest including world and American history, sports, movies, geography, the space program, actors, actresses, the Bible, church history, literature, classical music, Shakespeare, etc.

He knew all the area codes in the country along with major city zip codes and all the television stations in the United States.

His friends called him "Kim-puter."

Kim did not walk until he was four years old and continued to have problems with hand-eye coordination and balance. Even as an adult his father had to help Kim bathe, dress, button his shirts, brush his teeth and comb his hair. That kind of assistance was essentially a 24/7 responsibility that his father patiently carried on throughout Kim's life.

Kim Peek was the inspiration for the movie **Rain Man.**

Savants (formerly known as idiot savants) often have low IQs and lack the education but in some cases demonstrate abilities that would be remarkable for someone with a genius IQ.

Psychologist Joseph Chilton Pearce wrote about savant syndrome in his book *Evolution's End: Claiming the Potential of Our Intelligence*. In "*Chapter 1: Idiot-Enigma*" he said that savants are untrained and untrainable, illiterate and uneducable. Few can read or write. They have apparently unlimited access to a particular field of knowledge that we know they cannot have acquired. Ask mathematical savants how they get their answer and they will smile, pleased that we are impressed but unable to grasp the implications of such a question. They say that the answers just pop up in their mind.

One mathematical savant was shown a checkerboard with one grain of rice on the first of its 64 squares. He was then asked how many grains of rice there would be on the final square if the grains of rice were doubled on each square. Forty-five seconds later he gave the correct answer, which exceeds the total number of atoms in the Sun.

Lightning calculating is the rapid solution of complex multiplication or division problems. Often there is an innate facility with prime numbers and square roots among savants with no idea or explanation as to "how they do it."

Darold A. Treffert wrote in his book *Islands of Genius: The Bountiful Mind of the Autistic, Acquired, and Sudden Savant*: "Frequently this extensive calculating ability is seen with complete absence of other very simple arithmetic skills."

"In the movie Rain Man, for example, Raymond Babbitt is able to instantly multiply numbers in his head (4343 x 1234 = 5,359,262) and also give the correct answer for computing the square root of 2130 (46.15192304). Yet when asked how much money he would have left from a dollar if he spent 50 cents he says 'about 70.' Asked how much a candy bar would cost he says 'about a hundred dollars.' Asked how much a sport car would cost, his answer was the same, 'About a hundred dollars.'"

The idea that mind and memory are in the brain and in the System provides a model for how savants produce mathematics, calendaring calculations, music, and other information.

Chapter 7
Man of Future

In the not so distant future, human society will be divided into two major groups: natural people and posthumans. There will be two groups of posthumans: enhanced people (technologically enhanced humans in biological bodies) and digitized people (individuals whose minds were uploaded into computers or robotic bodies). The digitized people will be immortal and incredibly intelligent, which will result in owning huge financial and other resources. Man will be a mortal animal no more. The digitized people will live not only in computers and sophisticated virtual realities but they can also live in robotic bodies, various smart machines, and biological bodies especially adapted for the purpose. In these biological bodies the digitized individuals could enjoy the life of the past: having a drink with friends, dinner with the family, visiting some nice place, safely taking part in wild parties (Roman-style orgies or futuristic psychedelic jamborees), etc.

Mind uploading is not possible in the near future because part of the mind and the memories are not in the brain but in the System.

In the brain is the primitive part of our mind. The higher part is in the System.

We don't know if it will be possible to copy the part of the mind and the memories that are not in the brain in order to copy and upload the whole personality of a given individual. Future generations will do their best to make a copy of the mind and the memories of people. I think that they will successfully resolve that problem. But, on the other hand, they will not get the tremendously useful intellectual and creative assistance by the super-intelligent and all-knowing System. They will lose the advantage of being part of the System which is introducing new knowledge into the society. All new discoveries are made in cooperation with the System. The collaboration with the System is very important because future humans have to compete successfully with other space civilizations in order to survive. They will need advanced science, superior technology, and super weapons. How the future humans will resolve this problem we don't know.

Many researchers are convinced that biological intelligence is only a transitory phenomenon, a fleeting phase in the evolution of intelligence in the universe. Man is not the final product of evolution.

It is possible that all biological and artificial and still unknown forms of intelligences (product of the far future) in our Universe are just precursors for the future single intelligence.

We still live in the animal period of the evolving intelligences.

The digitized people will be tremendously intelligent, for they will think a million times faster than present-day humans because electronic circuits communicate much faster than neurons do and the data processing will be organized differently than in our brain. And thanks to the quantum and other novel data processing technologies, they will have a superb memory and instant access to all knowledge of humanity and will constantly self-improve extremely fast. The digitized mind of these people will be coupled with personal assisting Artificial Intelligence, resulting in a prodigious intelligence level. It is possible that the digitized consciousness and the personal AI will merge into a single mind.

The humans also will create a unique aide, partner, and possibly a dangerous competitor in their pursuit for excellence—autonomous Artificial Intelligence.

There will be a huge intelligence explosion on Earth.

The future people and AIs will develop science and technology at incredible speeds reaching technological singularity—this is the point beyond which the changes to science, technology, intelligences, and human civilization become utterly incomprehensible and unforeseeable to us.

Some researchers suppose that technological singularity is the moment when machines reach a level of intelligence that exceeds that of humans, and eventually they will annihilate humankind. This is a wrong assumption because digitized people will have the same technological resources and data processing capacity as machine intelligence.

The machine bodies and the minds of the digitized people will become so advanced that they will become a new form of existence which is now totally incomprehensible to us.

The data processing technologies will become so sophisticated that posthumans would create and own very complex virtual realities (simulated worlds) populated by artificial intelligent creatures not knowing that they are simulations.

After a very long evolution, the minds of the posthumans (digitized minds in various machine and biological bodies) will be interconnected and will form a sort of collective consciousness—one mind in numerous bodies—becoming one immortal, superfast, all-knowing omnipotential, supermind/creature. Is this the ultimate goal of the evolution of this Universe? Why is the System (the superior intelligence) producing new superior intelligence that way? Do they have better ways to produce new superior intelligences? Is this the only way to produce superior intelligences? Maybe with every evolutionary cycle the Universe is producing a better superior intelligence? We still cannot answer these questions. But reaching the so-called Omega Point (the final point to which all of history is progressing) means no end of the evolution. The new super intelligences enter a world populated with a great number of superior intelligences.

The Omega Point is a term coined by the French Jesuit Pierre Diehard de Chardin. It describes a maximum level of complexity and consciousness toward which he believed the Universe was evolving.

The languages of the distant future will be so different than the languages of our time that even an interpreter could not translate a text or conversation from some tongue of that future into a contemporary language because the science, technology, arts, culture, environment, and the common things will be wildly different, advanced, sophisticated, and complicated. The world and the concepts explaining that world will be incomprehensible to us. The contemporary vocabulary is insufficient for translating the future language or the language of a contemporary superior intelligence.

The posthumans will be incredibly intelligent, will possess knowledge that contemporary people can't comprehend, and will think and communicate via electronic signals millions of times faster than us, exchanging a tremendous volume of information every single second. We cannot follow such conversation. There will be new ways of thinking and communication totally incomprehensible to us.

We cannot comprehend the future and the languages of the future.

English is not the language of the distant future. Neither are any other contemporary languages on Earth. They all will be dead languages. These are the languages of the speaking animals with early science and budding technology on their way to the incomprehensible future.

World languages come and go. Sumerian, Sanskrit, Akkadian, Egyptian, and

Latin are dead languages, even though they played a significant role at the time of their triumph.

This book is written in the dead English language—from the perspective of the future.

Most probably, the owner of our Universe and the operators of the System controlling our world are so advanced, super-intelligent, and sophisticated that we can't understand what they are.

To us, our Universe seems tremendously huge. But in the world of the originator it could be just a gadget, the best little gadget around for the price.

To us, our Universe seems extremely old—it probably originated 13.8 billion years ago. Such an enormously long time period could be inappropriate to the owner of the Universe. But time could flow differently in our Universe and outside of it. One million years in our world could correspond to one hour in the world of the owner of the Universe. On the other hand, his thinking could be millions of times faster than ours and he could see and analyze every little detail of our history that is represented in just one hour of his time (one million years our time).

Why did the originator of our world create us? Or maybe the owner just bought our Universe, precisely as we buy a formicarium (ant farm) filled with ants.

The future people (posthumans in whatever physical form they are) will start creating worlds like our Universe, with a legion of intelligences on innumerable planets. Why would they create such simulations? Probably for the same reason people are creating video games and scientific simulations now—for fun, for profit, for research. Perhaps this could be the answer to the question of why our Universe was created.

We still don't know what could be the end purpose of the Universe, the life (natural and artificial in the near future), and the intelligences (natural and artificial). One of the main characteristics of the Universe, life, and intelligences is evolution—their gradual development into more complex forms. What could be the end result of the development of the intelligences in our Universe: a great number of super-civilizations with innumerable biological creatures, artificial super-beings, and machine intelligences or a single supermind?

It is also possible that the real purpose of the evolution of our Universe could be the development of the System into a higher form of intelligence and existence. All evolving life and intelligences (including humans) could be just transitory educational toys of the System.

Chapter 8
The Simulation Guides the Animals, Too

The System controls and guides not only humans but also the animals.

There are many cases of the amazing reuniting of man and dog. Sometimes they find their way home or to their masters across hundreds of miles or even thousands miles of unfamiliar terrain.

In 1929 Captain A. H. Trapman published in his book *The Dog: Man's Best Friend. A Book For All Dog Lovers*, the story of the amazing Prince, half Irish terrier, half Collie.

During World War I Private James Brown went to France with the 1st North Staffordshire Regiment in September 1914. His wife and small dog, Prince, were living in London. After a period of service Brown was granted leave to visit his family. After the furlough ended Prince was disconsolate and refused all food. Then the dog disappeared. The wife tried to trace him, to no avail. After ten days she sent a letter to her husband with the sad news that Prince was lost.

The wife was amazed when she received a letter from her husband saying that Prince had joined him in the trenches in France. Somehow the small dog had made his way through the streets of London, 70 miles of English soil, across the English Channel, traveled over 60 miles in France, and then found his master among an army of half a million Englishmen "…despite the fact that the last mile or so of intervening ground was reeking with bursting shells, many of them charged with tear-gas."

News of the amazing reuniting of man and dog spread quickly and the following morning Private Brown had orders to parade with Prince before the Regiment's Commanding Officer to verify the story.

The regiment adopted Prince as a mascot and he remained in France during the war. He was given a jacket made from an old khaki tunic and had his own identification disc.

When the amazing story of Prince was first told in the press, many people refused to believe it. The Royal Society for Cruelty to Animals investigated and proved its veracity.

The System successfully guided the small dog from London across the English Channel to the battle-field in France.

Animal migration is the relatively long-distance movement of individual animals, usually on a seasonal basis. It is found in all major animal groups, including birds, mammals, fish, reptiles, amphibians, insects, and crustaceans. The trigger for the migration may be local climate, local availability of food, the season of the year or for mating reasons.

Many mechanisms have been proposed for animal navigation: remembered landmarks, orientation by the night sky, orientation by magnetic field, olfactory navigation (animals build and remember a mental map of the odors in their area), orientation by the Sun, and so on.

In his book *Dogs That Know When Their Owners Are Coming Home* Rupert Sheldrake wrote that all attempts to explain the navigational ability of animals in terms of known senses and physical forces have proved unsuccessful.

Charles Darwin proposed that birds remember the twists and turns of the outward journey. Sheldrake wrote that it was refuted by taking pigeons to an unfamiliar point of release in dark vans, within rotating containers, by devious routes. Some were even anesthetized throughout the journey. But when the birds were released, they flew straight home.

He wrote that the theory that they rely on familiar landmarks has also been ruled out because they can also return from unfamiliar places hundreds of miles away where they cannot see any recognizable landmarks.

In experiments carried out in the 1970s, pigeons were even temporarily blinded by being fitted with frosted-glass contact lenses. They still found their way home over great distances.

The hypothesis that the birds follow the Sun fails to explain how birds find their home with frosted contact lenses on, and at night, and in heavily clouded conditions.

A theory suggests that pigeons and other birds smell their home from hundreds of miles away. But even when the wind is blowing in the wrong direction, they find their home. Pigeons could still find their way home even if their nostrils were blocked up with wax, their olfactory nerves severed, or their olfactory mucosa anesthetized.

Navigation by the magnetic field has been tested by attaching magnets to pigeons. These should confuse their magnetic sense, but birds with magnets attached easily find their home.

Sheldrake proposed a different animal navigation mechanism.

"The idea of morphic fields linking animals to other members of their social group provides a basis for understanding both telepathic communication and a directional pull toward animals or people. The idea of morphic fields linking animals to particular places provides a basis for understanding the sense of direction as expressed in homing and migration. Thus the morphic field hypothesis may account for a wide range of unexplained powers of animals, both telepathic and directional."

Rupert Sheldrake wrote in his book *A New Science of Life* that there is a lot of circumstantial evidence for morphic resonance. In a long series of tests that started in Harvard in the 1920s and continued over several decades, rats were taught to escape from a water maze and subsequent generations learned faster and faster. The rats had learned to escape more than 10 times quicker at Harvard. When rats were tested in Edinburgh, Scotland and in Melbourne, Australia they started more or less where the Harvard rats left off. According to Sheldrake's theory, the Edinburgh and Melbourne rats received the new knowledge through morphic resonance.

Whether the System uses the so-called morphic fields and morphic resonance we still don't know, but we know for sure that similar controlling mechanism exists. Future scientists will prove the physics behind the control over the matter and minds of intelligent creatures and animals. Everything in our Universe is under the tight and inescapable control of the System.

Chapter 9
The Omnipotent Simulation Plays Games

A team at the University of Minnesota studied 1,400 pairs of identical twins. They found that separated identical twins showed remarkable similarities in a variety of characteristics.

In 1979, Jim Lewis met Jim Springer, and some startling similarities came out.

They were twins separated at four weeks old, who had grown up about 45 miles from each other and ended up leading almost identical lives. Both were apart for thirty-nine years.

Both were named Jim (James) by their adoptive families. Each grew up with an adopted brother named Larry. They each married a woman named Linda, divorced, and then married another woman named Betty. Both twins had previously owned a dog named Toy. Both preferred Miller Lite beer and chain-smoked Salem cigarettes. They both had sons named James—James Allen and James Alan.

They drove Chevrolets, enjoyed carpentry, and had similar basement workshops. Both disliked baseball, and relished stock-car racing. They had been firemen and sheriffs. Both chewed their fingernails down to a nub.

Both had been lackluster students in high school; their favorite subject was mathematics and their least favorite, spelling.

Both had the only house on the block with a white bench around a tree in the yard, built by themselves during the time right before they met.

Each twin had vacationed in Florida at the same three-block-long beach.

Each had had a vasectomy. Both left love notes to their wives throughout their houses.

They had voted identically in the previous three presidential elections.

The Jim twins died from the same illness on the same day.

So many shared life details between the Jim twins defied the odds of chance. Our knowledge of genetics and environment can't explain the Jim twins' conundrum.

The only rational explanation here is that their lives were controlled by some external agency which plays games with people.

The lives of all people are guided.

Intelligent life on Earth looks like a well-orchestrated puppet show.

The same body but the personality was changed with another one: is it possible?

Iris Farczády, a well-educated Hungarian girl, practiced as a spiritualist medium. She regularly became possessed by spirits, some of whom remained in control after the séances were over. In August 1933, when she was sixteen, the spirit of the 41-year-old Spanish charwoman Lucía Altarez de Salvio took control but did not leave, as previous entities had done.

Now Iris spoke only Spanish, a language she had apparently never learned nor had the opportunity to acquire. Lucía (Iris) understood no Hungarian.

Lucía's mastery of Spanish in the Madrilene dialect was precise. She liked to dance Flamenco and other Spanish or Gypsy dances. Her performance of the Flamenco can't be learned by watching it because of its difficult choreography.

Iris's family was not happy about the replacement of their daughter by an uneducated foreign woman who spoke only Spanish and whose interests now were cooking, cleaning, washing dishes, and performing Spanish songs and dances.

The case was reported in the magazine *Journal of the Society for Psychical Research,* April 2005, under the title "The case of Iris Farczady -

A stolen life" by Mary Rose Barrington, Peter Mulacz, and Titus Rivas.

Some researchers suggest that the case of Iris Farczády is a proof of survival after death: the spirit of the dead Lucía entered the body of Iris.

But it is most likely that the System not only totally controls the minds of people and molds their personality, but it can also change the personality of a given body with a new one.

The System can even make you speak a language you don't know, an ability sometimes reported during religious events, medium séances, UFO events, some paranormal phenomena, and after head traumas.

The Lucía/Iris example is proof that the reincarnation cases are just games of the System to convince people that they have eternal souls, and they will live forever incarnation after incarnation.

The headline of the *The New York Times* for October 9, 1910 said, ILLITERATE MAN BECOMES A DOCTOR WHEN HYPNOTIZED—Strange Power Shown by Edgar Cayce Puzzles Physicians.

"The medical fraternity of the country is taking a lively interest in the strange power said to be possessed by Edgar Cayce of Hopkinsville,

Ky., to diagnose difficult diseases while in a semi-conscious state, though he has not the slightest knowledge of medicine when not in this condition."

Edgar Cayce, nicknamed The Sleeping Prophet by his biographer Jess Stearn, was an American trance medium who claimed to channel his own higher self. Cayce's channeling sessions happened in a trance state that he would induce with help from his friend Al Layne or some other assistant. He had no memory of the trance experience.

Cayce was a backwoods Kentucky farm boy with an eighth grade education, but later during trance sessions, he would answer questions on subjects as varied as healing, how to perform a surgical procedure, events in the distant past, nutrition, lost items, future events, and of course some inevitable fancy New Age stuff so typical for the séances of all mediums—reincarnation, higher self, past lives of the individuals, angels, spirituality, Atlantis, spiritual vibration frequency, the secret of the Sphinx, etc.

When using channeled information one should know that useful information always goes along with lots of wild trash. Get used to it if you are going to explore the phenomenon.

Cayce gave over sixteen thousand trance readings, some lasting thirty minutes to an hour. He was successfully diagnosing and prescribing medications or herbs to sick people. Cayce didn't speak in vague terms but used precise medical terminology well beyond his education and training. In the course of his forty-one-year career, he reportedly saved thousands of people from all sort of diseases and crippling injuries.

Some séances were witnessed by university professors, church leaders, scientists, physicians, inventors, and other professionals. One of them was the prominent psychologist Hugo Münsterberg of Harvard Medical School. The famous magician Harry Houdini failed to debunk or explain the Cayce phenomenon.

In his book *There is a River* Thomas Sugrue wrote about the life of Edgar Cayce. The following story demonstrates that the readings of the prominent medium are not some sort of freak imagination as many hard-boiled skeptics suggest but reflect the real world. And second, the System knows every minute detail of life on our planet.

Dr. Ketchum said that in one reading he conducted a preparation was given called "Oil of Smoke." But it was not listed in the pharmaceutical catalogues. He took another reading and asked where it could be found. The name of a drugstore in Louisville was given. He wired there, asking for the preparation. The manager wired back saying he did not have it and had never heard of it.

"We took a third reading. This time a shelf in the back of the Louisville drugstore was named. There, behind another preparation—which was named—would be found a bottle of 'Oil of Smoke,' so the reading said. I wired the information to the manager of the Louisville store. He wired me back, 'Found it.' The bottle arrived in a few days. It was old. The label was faded. The company which put it up had gone out of business. But it was just what he said it was, 'Oil of Smoke.'"

Sidney Kirkpatrick wrote his book *Edgar Cayce: An American Prophet*, "At around the same time that Andrews first wrote to them, friends of Layne's made a trip to France to see the Paris Exposition. Among them were two of Edgar Cayce's most staunch critics. As a test, Layne asked them to keep a written record of their activities while at the exposition. Upon their return, Layne requested a trance session to describe their itinerary. Layne then took the subsequent reading and reported to his friends that on a certain day and hour they had visited the Art Building at the exposition and admired particular paintings, which he described in detail. Far more impressive—and quite embarrassing for the Paris travelers—Layne went on to recount Cayce's report on the details of an evening when they visited a striptease show."

After a complaint from the school teacher, Cayce's father ruthlessly tested him for spelling, eventually knocking him out of his chair with exasperation. At that point, Cayce "heard" a female voice in his head. She told him that if he could sleep a little "they" could help him, but he should put his head on the spelling textbook. The boy begged for a rest and put his head on the spelling book. When his father came back into the room and woke him up, he knew all the answers. In fact, he could repeat anything in the book. His father thought he had been fooling before and knocked him out of the chair again. Eventually, Cayce used all his school books that way.

Obviously the System can read textbooks and can implant the texts in the mind of the individuals. According to Cayce, later when school required not only good memory but also some thinking, he was again in trouble. When he was already famous, he said, "I'm the dumbest man in Christian County, when I'm awake."

Sidney Kirkpatrick described in his book *Edgar Cayce: An American Prophet* an innovative therapy recommended by Cayce during a trance reading.

The reading that most convinced Ketchum of the potential of the Cayce readings was for George Dalton, the wealthy owner of Hopkinsville's brickworks. Dalton—who weighed well over two hundred pounds—had broken his right leg both below and above the knee. Hopkinsville's other doctors said that Dalton would never walk again and that amputation would be necessary. But Ketchum—on trance advice from Cayce—said that the knee could be healed.

The subsequent reading recommended that Ketchum bore holes in the kneecap and leg bones, insert nails into them, and put Dalton in traction. Ketchum was dubious, at best. Inserting metal screws or nails into bone was a procedure that had never before been performed in Kentucky or anywhere else in the United States. However, there was no harm in trying the procedure. The worst-case scenario was that Dalton would lose his leg, which is what the other physicians anticipated from the start. Ketchum had nails made to Cayce's specifications. Assisted by another doctor and two nurses, he bore holes in the knee and leg bones and then inserted the nails. Two months later Dalton was back on his feet. The nails were still in his leg seventeen years later when he died.

What was the source of this innovative therapy which was well beyond the mental capacity, education, and medical expertise of Cayce?

Who or what made the precise diagnosis, not asking the patient a single question about his condition, offered healing recommendation, and was prescribing drugs while he was in trance? The mystical source used language that wasn't in Edgar's regular vocabulary. According to Cayce, this was some mystical higher self. He channeled his own higher self.

What could be this supposed higher self? There are a lot of books and articles giving recommendations on how to connect with the mysterious all-knowing higher self, but no author can say exactly what it is. They say something like that: the higher self is generally regarded as a form of being only to be recognized in a union with a divine source; the higher self is a part of an individual's metaphysical identity; others teach that the higher self is essentially our tie to the heavens.

Well, this is just one of many disguises of the System.

Later experiments done by Layne and Cayce would suggest that it wasn't only Cayce's higher self doing the diagnosing and prescribing drugs, but that his higher self was the conduit or channel for someone or something else.

Dr. Ketchum became aware that the source or higher self had a distinct "personality." The all-knowing mysterious source could also be abrupt and disliked what it considered inane questions. And sometimes it demonstrated a sense of humor. When a patient asked if a medication should be rubbed on the outside, he was simply told, "You can't rub it on the inside!" That not always pleasant sense of humor is well known to all people in contact with the System. Sometimes it even insults the channeling person and the people present.

Obviously, the source of information in not just a giant database, but there is a witty intelligence behind it that can mock people and sometimes even make practical jokes.

John Keel wrote in his book *The Mothman Prophecies* that he received a lot of information from many psychics and often their predictions came true.

He was convinced that Pope Paul was about to be killed at the Istanbul airport and there would be a blackout in New York. John rented a car, loaded it with food, bottled water, flashlights, and candles, and drove out to the Mount Misery area to await the predicted three-day blackout. On his way he stopped to see one of his contactees and was informed that an extraterrestrial spaceman had just been to see him and had left a strange message: "Tell John we'll meet with him later and help him drink all that water."

John Keel wasted months playing the mischievous games of the ufonauts (actually of the System), searching for nonexistent UFO bases.

Through the channel you don't get access to the immense wisdom of the superior mind in control of the Universe. Yes, you are contacting a source of immense wisdom but the System often is playing with you. Man is still too small for the System or the superior intelligence to make contact with him as an equal. We should never forget that along with some useful information, the channel offers a lot of trash and fancy stuff. It is teaching us.

Paco Rabanne, the fashion designer, parfumier, book author, clairvoyant, and multi-millionaire inventor of the all-metal miniskirt, predicted in his book *1999 Fire from Heaven*, published in May 1999, that during the solar eclipse of August 11, 1999, the Soviet space station Mir would fall like an avenging angel from the skies, laying waste to Paris and wreaking collateral havoc in the southwest of France. The point of impact would be the castle of Vincennes. Gigantic fires and plutonium debris would ravage the French capital and several southwestern cities.

He was so convinced in the accuracy of this channeled information that he had made arrangements to be absent from Paris and the southwest in early August.

Rabanne said, "My personal conviction is total. I'm aware I'm staking my honor and credibility on this and I'd like nothing better than to be proved wrong. Nonetheless, I'm giving all my staff who are not already on holiday a few days' leave. And my shops will be closed."

But the Soviet space station Mir didn't destroy Paris or any other city in the world, proving one more time that predictions only rarely come true.

In 1994, in his best-selling book *Has The Countdown Begun?*, Paco Rabanne predicted Armageddon sometime around 1996. Fortunately, that didn't happen either.

Rabanne believes he is 78,000 years old and comes from "the crystal planet in the constellation of the Eagle which orbits round the star Altair."

He has said he has had several lives. Paco Rabanne believes he was around Jesus in a previous life and has seen God three times, and has been visited by aliens.

Rabanne received a lot of useful information that helped him create his fashion empire, but he also received a grave lesson not to believe in everything he received from the channel.

Dr. Charles Laughead, a physician at Michigan State University, had been receiving messages through a medium, Dorothy Martin, a 54-year-old housewife. Extraterrestrial sources from the planet Clarion made a number of predictions, which all came true. Then the alien entities predicted that the world was going to end on December 21, 1954, because North America would split in two, the East Coast would sink into the sea, and great parts of Europe would be destroyed.

The entities said that "there will be much loss of life, practically all of it...It is an actual fact that the world is in a mess. But the Supreme Being is going to clean house by sinking all of the land masses as we know them now and raising the land masses from under the sea."

Only the chosen ones, a handful of earthlings, including Dr. Laughead and Dorothy, would be lifted up that night by extraterrestrial spacemen.

Dr. Laughead announced the news to the press and he was given wide coverage. With a group of believers, he awaited the catastrophic events, which, of course, didn't happen. Dr. Laughead lost his job and reputation.

The System never stops playing games.

Contact with the source of higher intelligence and superior knowledge does not mean that one will get direct access to the wisdom of the Universe. Mystery is a necessary ingredient in our lives. It stimulates our thinking, mental development, and science. We will get a lot of religious, environmental, social, and scientific mumbo-jumbo, practical jokes, and lies but in some cases also real knowledge on some future events, useful health and business advice, artists will get some help creating their works, scholars will "discover" new theories, etc. Humans should be able to understand and prove what part of this information is absurd and what is true. Human evolution and individual development always happen step by step. Science and the paranormal, the logical and the absurd, are intended to coexist side by side.

We have harsh teachers. Individuals have no value. Only the entire humanity has some temporary value. The ultimate goal of the masters is well beyond humans and humanity.

Billions of scientists in our Universe are discovering the same theories human scholars are finding out on Earth. Billions of alien Einsteins discovered the theory of relativity—actually, they got it from the System. A pretty humiliating idea! One of the many tasks of the controlling agency is to educate us, but it is also providing us with novel knowledge. Now sentient beings (biological or not) in our Universe are more like (bio)robots that are created, uplifted, organized, controlled, and educated by an external agency. You don't like the idea? I don't like it either, but I would prefer to accept the truth, instead of some self-misleading puffed-up belief about the great importance of human and alien creatures consciously exploring Nature to the full extent. I also like the notion that we are intelligent, self-governing, originative, independent, creative, and self-sustaining creatures with free will, but is it true?

Bitter facts are better than self-deluding illusions or doctrines if one is going to explore the world. Comforting lies have nothing to do with science.

We are living in a controlled environment and we are controlled evolving animals.

We should never forget that humans are under constant tight mind control and never trust what we see, hear, and think. Don't trust yourself. You are under control. Your mind is controlled. It is quite possible that your thoughts are not your own. You think and you are what the System decides.

The ancient Greek epic poem *Iliad*, traditionally attributed to Homer, is usually considered to have been written circa the 8th century BC.

The historical backdrop of the poem is the time of the Late Bronze Age, in the early 12th century BC.

In the poem the ancient Greek gods constantly spoke to the legendary heroes of the

Trojan War. The gods caused them to quarrel and fight and provided them with strategies.

Julian Jaynes, a professor of psychology at Princeton, wrote in his book *The Origin of Consciousness in the Breakdown of the Bicameral Mind*,

"Who then were these gods that pushed men about like robots and sang epics through their lips? They were voices whose speech and directions could be as distinctly heard by the Iliadic heroes as voices are heard by certain epileptic and schizophrenic patients, or just as Joan of Arc heard her voices…. The Trojan War was directed by hallucinations. And the soldiers who were so directed were not at all like us. They were noble automatons who knew not what they did."

The term bicameral mind coined by Jaynes is a controversial hypothesis in psychology and neuroscience which argues that the ancient human mind once was divided in two: one part of the brain (right hemisphere) was speaking and commanding, and the second part (left hemisphere) was listening and obeying. The experiences, advice, commands, and memories of the right hemisphere of the brain were transmitted to the left hemisphere via auditory hallucinations. Rather than making conscious decisions, the person would hallucinate a voice or "god," and would obey without question. The Iliadic heroes were not conscious as modern people are.

"The characters of the *Iliad* do not sit down and think out what to do. They have no conscious minds such as we say we have, and certainly no introspections."

Whenever a decision has to be made, a voice comes telling people what to do. "These voices are always and immediately obeyed. These voices are called gods. To me this is the origin of gods. I regard them as auditory hallucinations."

Jaynes suggested that voice-hearing was universal in ancient times. All early civilizations have been ruled by auditory hallucinations or gods.

The evolutionary breakdown of this division gave rise to consciousness in humans about 3,000 years ago.

The hypothesis that the human mind was divided into two parts and the first part was dictating to the other part of the individual what to do and the people obeyed actually means that part of the human mind-brain should be far more intelligent and knowledgeable than the other part. Quite amusing! The silly lower self of the mind receives advice, information, and commands from the mystical, all-knowing (including the future), godlike higher self. The second part that knows everything and is much more intelligent than all humans and that knows the future is actually the System! The human brain is not bicameral but the human mind is divided into two parts: one is in the brain, the second one is in the System.

Richard Dawkins in his book *The God Delusion* wrote about *The Origin of Consciousness in the Breakdown of the Bicameral Mind*: "It is one of those books that is either complete rubbish or a work of consummate genius, nothing in between! Probably the former, but I'm hedging my bets."

In ancient times probably the System was guiding people more directly, with direct commands through auditory and visual hallucinations, but it was a long time before the Trojan War.

Gods speaking to heroes during the Trojan War could be just a literary device of the ancient authors who make their heroes hear voices from gods because this makes the ordinary individual a hero. Only the chosen ones, the heroes, and the oracles are communicating with the gods. The rest are just part of the vulgar herd.

Now the control and the embedding of information into our minds is more subtle. Instead of direct auditory and visual hallucinatory commands, most contemporary people have intuition, dreams, subconsciousness, and only on some occasions mentally healthy individuals get controlled visual and auditory hallucinations. Sometimes it is difficult to say what is pathology, what is creative thinking, or what is direct information from the System.

John Nash, Nobel Prize laureate in Economic Sciences, spent several years at psychiatric hospitals being treated for schizophrenia. He was asked, "How could you, a mathematician, a man devoted to reason and logical proof…how could you believe that extraterrestrials are sending you messages? How could you believe that you are being recruited by aliens from outer space to save the world?" Nash said, "Because the ideas I had about the supernatural beings came to me the same way that my mathematical ideas did. So I took them seriously."

They are here and they control us—the flying saucers, the aliens from all sorts of planets and dimensions, the deities and their religious miracles, the paranormal phenomena, the fairies, the demons, and many other products of the System. They are the long hand of the superior civilization in charge. They are in our minds and lives. The superior intelligence is here and we can see its reflection in the phenomena it is displaying on Earth and in our minds.

Chapter 10
Creativity, Intuition, and The Genius

What is a genius and how does he make discoveries?

Researchers are often surprised that through contacts with religious entities, ghosts, aliens, spirits, etc., the contactees are able to answer questions that appear to be well beyond their mental capacity, education, and technical expertise, which proves one more time that they are not making it up, but that they receive information from an external source of higher intelligence and superior knowledge.

All great works (artistic, scientific, social, military, political, etc.) are done in cooperation with the System. Most great persons realize that they are collaborating with some higher intelligence they call by different names.

Without the System, the intellectual capacity of humans would be lower and our civilization would advance at a much slower pace.

Uplifting primitive races by a superior civilization is an ages-old idea very popular on our planet—ancient writings described all sorts of gods and visitors from the cosmos creating, uplifting, and teaching humans.

The System is assisting people in multiple disguises (muses, guiding spirits, angels, alien visitors, etc.) to create a civilization.

The novel *Uncle Tom's Cabin* by Harriet Beecher Stowe was published in 1852 and was an immediate success.

President Abraham Lincoln was quoted as saying when he met Stowe, "So you're the little woman who wrote the book that made this great war!"

Later she would say that she did not actually write it on her own. "I did not write it. God wrote it. I merely did his dictation."

Ancient Greeks believed that the poets were possessed by a daemon (deity or guiding spirit), which is why Plato distrusted poets—they were not ruled by reason. Homer opens *The Odyssey* with an invocation to the muse and asks for her guidance in telling the story.

The Greek philosopher Socrates claimed to have a daemon (a divine entity) that came in the form of a voice to warn him about mistakes and to speak wisdom to him. The voice offered him guidance in regard to everyday tasks.

Socrates claimed to have heard it since childhood.

John Milton said that his poem *Paradise Lost* was dictated to him, at night or in the early morning, by his "celestial patroness," the heavenly muse he called Urania.

Sir Herbert Read wrote in his article "The Poet and his Muse," "Blake is not a complicated case—he believed quite simply that in the act of writing poetry he was being dictated to by heavenly spirits, sometimes anonymous, sometimes recognizable as historical characters. There is not one 'celestial patroness,' but many 'angels' or 'authors' in Eternity."

The French writer Edmond de Goncourt wrote, "One does not write the books one wishes. There is a fatality about initial luck, which dictates its idea to us. Then an unknown force, a superior will, a sort of necessity to write, prescribe the work and bring you your pen, so much so that the book which comes from your hands does not seem to proceed from yourself. It astonishes you, as though it were something within you of which you had no knowledge."

Ray Bradbury noted, "Your intuition knows what to write, so get out of the way."

Robert Graves, English poet, translator, and novelist wrote, "Intuition is the supra-logic that cuts out all the routine processes of thought and leaps straight from the problem to the answer."

Jim Shepard, author and professor of creative writing, said, "It turns out that our intuition is a greater genius than we are."

David Lynch wrote, "Intuition is the key to everything, in painting, filmmaking, business—everything."

Melvin Morse and Paul Perry wrote in their book *Transformed By The Light* about the creative writing of the bestselling author David G.

At a particularly difficult point of the story, when he is mulling over his plot from every possible angle and can't find a way to proceed, the unseen hand of a guardian angel takes over and guides him through his work.

David says, "It is almost like automatic writing, I don't know how else to describe it. Sometime there are large sections of my writing that I don't remember having produced. Even my wife, who works closely with me, doesn't recognize this writing as being my own. She says, 'this doesn't look like your style,' and I have to agree with that. There are sections of my book that seem to have come to me from somewhere else."

There is a great number of channeled books on various subjects. The religious texts (scriptures) were also dictated by entities considered deities.

The huge volume titled *Oahspe: A New Bible* was written by the New York dentist Dr. John Newbrough in 1880. He had been a spiritualist and medium since the early 1870s.

Newbrough wrote publicly about how the *Oahspe* came about through automatic writing. He sat at a typewriter for half an hour each morning, at which time his hands would automatically type without his knowledge of what was being written. The drawings in the book contain symbols resembling hieroglyphs. Sometimes he had the urge to write and sat down with pen and paper until a bright light enveloped his fingers and they started to automatically write and draw.

The sophisticated language and the contents of the book are well beyond the expertise and the education of Newbrough.

The first (1882 edition) publication contained various glyphs, whose resemblance to real Egyptian hieroglyphs was attested to by Professor Thomas Ward, who claimed to have deciphered the hieroglyphics on the Cleopatra's Needle obelisk in Central Park. Ward was present at *Oahspe*'s first presentation, as was Dr. Cetliniski, an Oriental scholar, who affirmed that mere mortals could not have produced such a book and that "supernatural agents" must have been responsible.

A *Course in Miracles* is a book by Helen Schucman who claimed that the book had been dictated to her, word for word, via "inner dictation" from Christ. The book is considered to have been borrowed from various New Age movement literature. Actually, the New Age texts for this and many other books, articles, lectures, etc. were delivered by the System.

Jane Roberts described the process of writing the Seth books as entering a trance state. She said Seth would assume control of her body and speak through her, while her husband wrote down the words she spoke. They referred to such episodes as "readings" or "sessions."

Roberts claimed no authorship of these books beyond her role as a medium. This series of Seth books totaled ten volumes. The material is regarded as one of the cornerstones of New Age philosophy.

The Seth Material discusses a wide range of metaphysical concepts, including the nature of God referred to as "All That Is" and "The Multidimensional God"; the nature of physical reality; the origins of the Universe; the nature of the self and the higher self; the story of Christ; the evolution of the soul and all aspects of death and rebirth, including reincarnation and karma, past lives, after-death experiences, guardian spirits, and ascension to planes of higher consciousness; the purpose of life and the nature of good and evil; the purpose of suffering; multidimensional reality, parallel lives, transpersonal realms, etc.

Since Roberts' death, others have claimed to channel Seth.

UFO contactees, mediums, shamans, witch doctors, healers, medicine men, fortune-tellers, oracles, seers, soothsayers, savants, visionaries, psychics, priests, gurus, prophets, saints, mystics, holy ones, light workers, initiates, doctrine teachers, adepts, spiritual masters, kung-fu masters, prophets, tai chi masters, yoga masters, saints, sages, authors, artists, scientists, and many others communicate with the multiple disguises of the System which are alien robots, extraterrestrials, super-intelligent alien computers, alien spaceships, spirits of dead people, creatures from the etheric plane, transdimensional beings, all sort of gods and goddesses, mythical creatures, spiritual entities, magical animals, the Supreme Spiritual Being, the Supreme Intelligence, Higher Beings,

Confederation of Planets, the Saturn Tribunal (or Council), the Intergalactic Confederation, the Cosmic Consciousness, the Superintelligence from the future, the Infinite Creator, various energy beings, ancient sages, muses, angels, archangels, guiding spirits, ghosts, the devil, demons, the Higher Self, the Universal Mind, the Universal Consciousness, the Brotherhood of Light, the Source, the "energy personality essence no longer focused in physical matter" called Seth channeled by Jane Roberts, the Dr. Adolf Fritz' spirit channeled by the Brazilian psychic surgeon Arigo, the interdimensional being Ashtar which a number of people have channeled, and many more.

In his book *Evolution's End: Claiming the Potential of Our Intelligence*, psychologist Joseph Chilton Pearce wrote about an amazing event when he was in his early 30s. He was engrossed in theology and the psychology of Carl Jung. Pearce wrote that he was obsessed by the nature of the God-human relationship, and his reading on the subject was extensive. One morning his five-year-old son came into his room, sat down on the edge of the bed, and launched into a twenty-minute discourse of the nature of God and man.

Pearce wrote, "He spoke in perfect, publishable sentences, without pause or haste, and in a flat monotone. He used complex theological terminology and told me, it seemed, everything there was to know. As I listened, astonished, the hair rose on my neck; I felt goose bumps, and, finally, tears streamed down my face. I was in the midst of the uncanny, the inexplicable. My son's ride to kindergarten arrived, horn blowing, and he got up and left. I was unnerved and arrived late to my class. What I had heard was awesome but too vast and far beyond any concept I had had to that point. The gap was so great I could remember almost no details and little of the broad panorama he had presented. My son had no recollection of the event."

It is more than obvious that the five-year-old boy had absolutely no knowledge about such an intricate subject as theology. The external agency that took control over the mind of the little boy in order to provide a sophisticated twenty-minute discourse of the nature of God and man was the System.

Under the heavy influence of Sigmund Freud and Carl Jung, contemporary science claims that thoughts that don't originate in the consciousness are a product of the unconscious part of the mind or the collective unconscious.

Popular among many researchers is the myth that the unconscious is the source of the intuition, creativity, the home of the alleged higher self, where the paranormal originates.

Carl Jung invented the collective unconscious to explain some phenomena of the human consciousness, but he didn't explain how the brains/minds of all people on Earth are connected so that they share a common unconscious mind.

In 1936, Jung delivered a lecture "The Concept of the Collective Unconscious" to the Abernethian Society at St. Bartholomew's Hospital in London. He said:

"My thesis then, is as follows: in addition to our immediate consciousness, which is of a thoroughly personal nature and which we believe to be the only empirical psyche (even if we tack on the personal unconscious as an appendix), there exists a second psychic system of a collective, universal, and impersonal nature which is identical in all individuals. This collective unconscious does not develop individually but is inherited. It consists of pre-existent forms, the archetypes, which can only become conscious secondarily and which give definite form to certain psychic contents."

How is the collective unconscious inherited? We don't know of such an inheritance mechanism. And second, the inheritance is individual, not collective. And how are the unconscious minds of all people on our planet connected?

The source of novel, theological, mythological, etc. information is the System. It delivers content to the human conscious mind and the collective unconscious.

Intuition is the ability to acquire knowledge without inference or the use of reason. It is a direct perception of fact, truth, hypothesis, theory, etc., independent of any reasoning process.

In cases of intuition, science suggests that there is a reasoning process that happens somewhere deep in the unconscious mind of the individuals. Sometimes this is true, but in many other cases this explanation fails to reveal how we get novel knowledge through intuition.

Melvin Morse and Paul Perry wrote in their book *Transformed By The Light: The Powerful Effect Of Near-death Experiences On People's Lives* about receiving new knowledge intuitively:

"Some claimed that they had archived higher intelligence after their NDEs. One, a snowplow operator in upstate New York with all-American name of Tom Sawyer, found himself writing a string of numbers and symbols within a year of his experience. He didn't know why he wrote them or what they meant, but he frequently found himself doodling during coffee breaks or in the evening after work. When showed these musings to a college professor he found that he was writing the equations of Max Plank, a physicist, who contributed much of what we know today about atomic theory. Sawyer now says the his NDE was a " short course in nuclear physics." But why and how did a person with high school education get such information?"

The equations of Max Plank were well beyond the education and technical expertise of the man who doodled them, which proves that they received information from an external source of higher intelligence and superior knowledge. They were not result of his unconscious mind.

When a healer is preparing a medicine, he/she usually uses expertise acquired from other healers, textbooks, or notes. But for some diseases they have to prepare a medicine not known to them. In such cases they hear in their head voices from gods, saints, deceased relatives, or dead healers, dictating the herbs and other ingredients they need and how exactly to use them. Some healers see how to prepare a new medicine while dreaming. Where does the information about which ingredients they need and how to prepare the new remedy come from? Scholars say that the healer intuitively knows this. This is not a satisfactory explanation because it does not answer the question of where the information comes from. Some scientists claim that the information comes from the subconscious, but how did this information get into the subconscious mind? It was not concocted there. It was received from somewhere. There is not some magic supercomputer that knows everything hidden somewhere in our unconscious mind.

The painter Paul Klee said about the work of the artist, "He neither serves nor rules, he transmits. His position is humble and the beauty at the crown is not his own. He is merely a channel."

Luiz Gasparetto was a Brazilian medium, painter, and TV presenter. He became well-known for mediumistic paintings attributed to many famous deceased artists. Gasparetto, with no training as an artist, usually worked in near darkness, at tremendous speed, and only occasionally looking at what he was painting.

Each painting had the unmistakable print of the channeled dead artist, who sometimes spoke through Gasparetto, offering healing counsel and advice to individuals attending the painting sessions.

He explained that the paintings had already been completed in the spirit world and were laid over his canvas like a template; he simply followed the designs. His movements were guided by the "thoughts" of the master who was painting through him.

Sometimes his eyes were tightly shut. Each painting took about 5 minutes. He produced thousands of pictures, sometimes done with both hands producing two different drawings by two different artists at the same time. When asked, "How do you do it?" Gasparetto answered, "I don't do it. They do."

Gasparetto called himself a spirit channeler, a medium through which many old masters chose to work.

The clairvoyant John Newbrough, who channeled the book *Oahspe,* could paint in total darkness with both hands at once.

Heinrich Nusslein, a German automatic artist, painted in total darkness, producing small pictures in three or four minutes.

Susannah Harris could complete an oil painting in two hours when blindfolded and painting it upside down.

The System has a perfect control over the human mind and body.

Michael Jackson said that the best songs write themselves. He believes that they were already up there before one is born.

Dave Sabo, an American musician best known as the guitarist of the heavy metal band Skid Row, stated,

"I am very aware that when I write a good song I am just acting like as a messenger. It comes from a higher source. I'm not so egotistical that I think I done this all on my own. I'm very, very aware there is a higher source. I am just the messenger."

Many of the great musicians merely waited for the song to appear in their heads or to hear it in their dreams.

John Lennon said that when really good music comes to him that has nothing to do with him because he is just the channel.

When people encounter the words *intuition* and *creativity*, most often they visualize an artist: painter, actor, piano player, singer, writer, music composer, sculptor, poet, and so on. What about logic-driven professionals like

Einstein, Steve Jobs, and other inventors, scholars, businessmen—do they believe and use intuition, or do they rely solely on logic and reason?

Steve Jobs said, "Intuition is a very powerful thing, more powerful than intellect, in my opinion. That's had a big impact on my work."

For Einstein, insight and scientific discoveries do not come from mathematics, analysis, or logic. They come from intuition.

Immanuel Kant wrote, "All human knowledge thus begins with intuitions."

The great inventor Thomas Edison stated, "People say I have created things. I have never created anything. I get impressions from the Universe at large and work them out, but I am only a plate on a record or a receiving apparatus—what you will. Thoughts are really impressions that we get from outside."

Rene Descartes wrote, "The two operations of our understanding, intuition, and deduction, on which alone we have said we must rely in the acquisition of knowledge."

The mathematician Henri Poincaré said, "It is through science that we prove, but through intuition that we discover."

Jonas Salk, an American medical researcher and virologist who discovered and developed the first successful polio vaccine, wrote, "Intuition will tell the thinking mind where to look next."

"It is always with excitement that I wake up in the morning wondering what my intuition will toss up to me, like gifts from the sea. I work with it and rely on it. It's my partner."

Jim Shepard, an American author and professor of creative writing and film said, "It turns out that our intuition is a greater genius than we are."

Carl Jung wrote, "In such doubtful matters, where you have to work as a pioneer, you must be able to put some trust in your intuition and follow your feeling even at the risk of going wrong."

Aristotle said, "Intuition is the source of scientific knowledge."

Alexis Carrel, a Nobel Prize laureate, wrote, "All great men are gifted with intuition. They know without reasoning or analysis, what they need to know."

"Intuition comes very close to clairvoyance; it appears to be the extrasensory perception of reality."

Nikola Tesla suggested, "My brain is only a receiver. In the Universe there is a core from which we obtain knowledge, strength, and inspiration. I have not penetrated into the secrets of this core, but I know that it exists."

Many great scientists, thinkers, researchers, and artists confirm that when it comes to novel knowledge, intuition is the source of new ideas. Humanity's science, technology, and arts are assisted by the System created by a superior intelligence.

Many scientists, engineers, people of art, authors, etc. received creative information in their dreams.

The aviation pioneer Igor Sikorsky reported a precognitive dream in his autobiography *The Story of the Winged – S*.

In 1900, he was about 11 years. Sikorsky had a dream and for several days he lived under the impression of that dream and always remembered the details.

"I saw myself walking along a narrow, luxuriously decorated passageway. On both sides were walnut doors, similar to the state room doors of a steamer. The floor was covered with an attractive carpet. A spherical electric light from the ceiling produced a pleasant bluish illumination. Walking slowly, I felt a slight vibration under my feet and was not surprised to find that the feeling was different from that experienced on a steamer or on a railroad train. I took this for granted because in my dream I knew that I was on board a large flying ship in the air."

"I was always interested in flying - I dreamed about it even when I was a small boy. However, at that time [1900] flying was considered completely impossible. The very expression of 'he was building a flying machine' was considered equivalent to saying that the man was crazy."

Sikorsky became an aircraft designer. About three decades after the dream in 1900, he went aboard one of his own four-engine Clippers to inspect a job of interior decorating done by Pan American Airways. He immediately recognized the cabin as identical to the one in his dream decades ago.

The interior of the future Sikorsky's aircraft was known in 1900, about 30 years before the inventions of the four-engine Clippers. The engineers and decorators just copied the design they received by the System in their minds.

The future is simulated and controlled by the System. Even such small details as future fashion is predetermined. The fashion to come is not invented by famous designers but by the System. UFO witnesses reported that some aliens or men in black were wearing clothes that would come into fashion years later.

Srinivasa Ramanujan (1887-1920) was an Indian mathematician. Though he had almost no formal training in pure mathematics, he made substantial contributions to mathematical analysis, number theory, infinite series, and continued fractions, including solutions to mathematical problems then considered unsolvable. Ramanujan initially developed his own mathematical research in isolation. Seeking mathematicians who could better understand his work, in 1913 he began corresponding by mail with the English mathematician G. H. Hardy at the University of Cambridge, England. Recognizing Ramanujan's work as extraordinary, Hardy arranged for him to travel to Cambridge.

During his short life, Ramanujan independently compiled nearly 3,900 results (mostly equations). Many were completely novel. His original and highly unconventional results have opened entire new areas of work and inspired a vast amount of further research.

Ramanujan credited his substantial mathematical capacities to divinity and said the mathematical knowledge he displayed was revealed to him by his family goddess Namagiri

Thayar, who showed him in dreams the equations of his formulas.

In the 1865, the German chemist August Kekule published a paper on the hexagonal ring structure of the six carbon atoms in the benzene molecule.

He said about his discovery, "[One evening] I was sitting there, working on my textbook, but it was not going well; my thoughts were on other matters. I turned my chair towards the fireplace and sank into half-sleep. Again the atoms fluttered before my eyes. This time smaller groups remained modestly in the background. My mental eye, sharpened by repeated visions of a similar kind, now distinguished larger forms in a variety of combinations. Long chains, often combined in a denser fashion; everything in motion, twisting and turning like snakes. But look, what was that?! One of the snakes had seized its own tail, and the figure whirled mockingly before my eyes. I awoke in a flash, and this time, too, I spent the rest of the night working out the consequences of the hypothesis."

Kekule said: "Gentlemen, let us learn to dream, and perhaps then we will find the truth . . . but let us also beware not to publish our dreams until they have been examined by the wakened mind."

Another prominent chemist, Dmitri Mendeleev, who created the periodic table, used it to correct the properties of some already discovered elements, and also predicted the properties of elements yet to be discovered, made his discovery in similar circumstances, "I saw in a dream a table where all elements fell into place as required. Awakening, I immediately wrote it down on a piece of paper, only in one place did a correction later seem necessary."

Kekule, Mendeleev, and many other scientists, artists, etc. made their discoveries, saw their pictures, heard their music, etc. in their dreams. The researchers suggest that their discoveries are the result of their unconscious intellectual activity. In some cases this is so, but in other cases the System implanted the new discovery. We can't know what the source is of the new finding—the unconscious intellectual activity or the System. It's much more comforting to think that we have made the discovery, but in most cases the discovery is a result of the implanting of new findings in our mind by the System.

In his book *Genius: The Natural History of Creativity* the psychologist Hans Eysenck wrote:

"Two notions, ideas, concepts - call them what you like - have always been attached to the problem of creativity. It has been widely surmised that the creative genius generates his major ideas by way of intuition, rather than rational thinking; reason can test and prove or disprove the insights achieved by intuition, but cannot produce them. Furthermore, the process by means of which intuition works is unconscious; the Unconscious, whether with or without a capital 'U', is the cradle of creativity."

We should not forget that the master of the human Unconscious is the System not the conscious mind.

The psychologist Wilhelm Wundt wrote:

"Our mind is so fortunately equipped, that it brings up the most important bases for our thoughts without our having the least knowledge of this work of elaboration. Only the results of it become conscious. This unconscious mind is for us like an unknown being who creates and produces for us and finally throws the ripe fruits in our lap."

For Wundt and many other psychologists, the unconscious mind is much more intelligent and creative than the silly conscious mind because it can produce new science and art.

Lewis Terman was an American psychologist best known for initiating the longitudinal study of children with high IQs called the Genetic Studies of Genius. Today it is known as the Terman Study of the Gifted, probably because Terman couldn't find a single genius but he rejected two kids who later became Nobel Prize winners. It was begun by Terman at Stanford University in 1921 to examine the development and characteristics of gifted children into adulthood. He began an ambitious search for the brightest kids in California, administering IQ tests to several thousand children. Those scoring above an IQ of 135 (approximately the top 1 percent of scores) were tracked for further study. Terman found 1,528 gifted kids—their average IQ was 151, with 77 claiming IQs between 177 and 200. The average IQ is 100. Individuals with an IQ above 140 are considered near genius or genius.

Ironically, the two future Nobelists Luis Alvarez and William Shockley were rejected by Terman for not being intelligent enough. William Shockley, the cocreator of the transistor, received the Nobel Prize in physics. Luis Alvarez also was awarded the Nobel Prize in physics. He became very popular with the hypothesis that the mass extinction of the dinosaurs and many other creatures during the

Cretaceous–Paleogene extinction event 66 million years ago was caused by the impact of a large asteroid on the Earth.

Ironically, not one of the more than 1,528 children with very high IQs received so high an honor as adults as Shockley and Alvarez. None of them grew up to become what many people would consider geniuses.

Most of the people with world's highest IQ seem to have ordinary jobs and ordinary lives. There are a lot of unremarkable people with a high or even extra high IQ.

So what is the secret to becoming a genius?

In ancient Rome, the genius was the guiding spirit of a person. The ancient people knew that their creative power came from an external agency.

To become a genius you need high enough IQ and to work in cooperation with the System. Without the System's help, one is just a smart kid who becomes a successful average man.

Arthur Schopenhauer wrote: "Talent hits a target no one else can hit. Genius hits a target no one else can see."

Many people have a higher IQ than Einstein but they are not geniuses.

Richard Feynman received the Nobel Prize in Physics. He gave a talk at his former high school in New York and told the students that when he took an IQ test at school, he scored 125.

The high achievements in science, arts, politics, etc. are a product of close cooperation with the System. Not all geniuses and high-profile professionals in all walks of life understand that they are working together with a higher intelligence. Some of them are aware that there is some sort of cooperation but they think this is God, some sort of angel, a divine guiding spirit, their higher self, or some inner guiding spirit part of their unconscious. Some individuals understand that they are receiving help from a higher intelligence but they are hiding that teamwork from the public. They are hiding the fact of cooperation very carefully so as not be ridiculed or to jeopardize their positions and careers, and not to be ostracized from society.

There are a great many books promising to make you a genius. They can't. Neither parents nor educators can do that. Only the System can make you a genius.

If possible, at some point in their evolution, people could try to cut the umbilical cord between developing humanity and the System. This is an important moment of civilizations' evolution because they will no longer receive useful scientific and other information from the external agency that turned the animals into intelligent creatures, creating a sophisticated civilization. Humanity is still too primitive to cut the umbilical cord connecting the System and people. If this happens now, we will all turn intellectually into kids and our civilization as we know it will collapse. We will relapse into a primitive agrarian society.

A theater producer protagonist from *Cyril and the Broadway Musical*, a movie from the *Jeeves and Wooster* TV show series based on the novels of P. G. Wodehouse, said that he has a ten-year-old son he relies on to tell him what Broadway audiences is gonna like or not gonna like. He says their mental age is about the same.

Chapter 11
Psychokinesis Parties

Wernher von Braun, the leading figure in the development of rocket technology in Nazi Germany and father of the U.S. space program, met Uri Geller and announced, "Geller has bent my ring in the palm of my hand without ever touching it. I have no scientific explanation for the phenomenon."

Many phenomena can be explained as control over people's mind by an advanced intelligence, but some riddles require control over matter. One such puzzle is metal bending.

From 1981 to 2005, aeronautical engineer Jack Houck hosted PK Parties (psychokinesis parties) with over 19,000 documented participants.

Michael Crichton wrote about his experience with spoon bending at the PK Party in his book *Travels*.

He said that there were about a hundred people there, mainly families with kids. They threw the silverware they had brought into the center of the floor.

Jack Houck told them to take spoons and shout, "Bend! Bend!"

People were pretty skeptical.

People were selecting spoons and shouting, "Bend! Bend!"

Michael Crichton bent several spoons and forks. He wrote:

"Of course, spoon bending has been the focus of long-standing controversy. Uri Geller, an Israeli magician who claims psychic powers, often bends spoons, but other magicians, such as James Randi, claim that spoon bending isn't a psychic phenomenon at all, just a trick.

"But I had bent a spoon, and I knew it wasn't a trick. I looked around the room and saw little children, eight or nine years old, bending large metal bars. They weren't trying to trick anybody."

Bonnie Greenwell wrote about spoon bending in *The Kundalini Guide: A Companion For the Inward Journey*.

He has also participated in a Psi-party, in which a group of several hundred people were taught how to bend metal objects, such as spoons, by concentrated thought, and most people were able to do this within an hour.

Greenwell wrote, "The children present were particularly adept at this because they had no preconceived notions that it couldn't be done. I have witnessed similar practices used to make seeds sprout within minutes, and to bend metal bars. This is not spiritual awakening, but rather playing energy games, and it does not require kundalini arousal to accomplish it. However it demonstrates the potential of the underused human psyche."

Army Colonel John B. Alexander wrote about his experience of metal bending in his book *Reality Denied: Firsthand Experiences with Things that Can't Happen - But Did*.

He considers Uri Geller a personal friend. During the time of their friendship, he observed him bend many objects. He was very impressed by one incident that took place in the U.S. Capitol Building. Geller was there to talk about the relations between Israel and the Soviet Union, but members of the audience insisted that he bend something.

There was no cutlery available in the room. Someone went outside and obtained a spoon from a guard's coffee cup.

Alexander wrote, "That is important as it proves that Uri had no means to prepare the object ahead of time.

"Obligingly, Geller held the spoon by the bowl with two fingers of his left hand. With his right forefinger, gently he stroked the neck of the spoon from above. Expecting this spoon to bend, I watched very carefully and noted that at no time did he touch both sides of the neck of spoon, as magicians do when faking the process. As Uri continued, the handle of the spoon visibly bent upward. Clearly, there was no force being applied that could account for the movement. Uri placed the spoon at the top of the back of a chair next to him. He then continued with his discourse about relations between Israel and the Soviet Union. As he did so, with an unobstructed view, I could tell the spoon was continuing to bend, even though Uri had no physical contact with it."

Who or what bent the spoon when Geller had no physical contact with it and was not paying attention to the spoon—he was talking about the relations between Israel and the Soviet Union? It was not him. Most probably the System is doing the bending of metal objects, the people are only active witnesses of the paranormal event. During psychokinesis parties ordinary people having no psychic abilities bend spoons and forks.

The British physicist John Hasted wrote about similar cases in his book *The Metal Benders*:

"During Geller's television performances other people both in the studio and in their homes would find that a latchkey held gently in the hand would bend of its own accord. In most West European countries, as well as Japan, South Africa and others, the television companies received letters and telephone calls reporting cutlery bending of its own accord in viewers' homes. Hundreds of such cases have been followed up in West Germany, and in Britain.

"In nearly all cases the effects began during or after the television performance."

Hasted reported that during a dinner at a hotel restaurant in Japan with Uri Geller, a spoon appeared to curve upwards gradually on the table, untouched by anyone, and in full view of all present.

During poltergeist manifestations, many people witnessed spoon and forks bending by themselves when not being touched by anyone.

There was an incident with the psychologist Carl Jung's knife. He organized spiritualistic séances in his youth, and in one of these a breadknife in a drawer inexplicably snapped into four parts, making a sound like a pistol shot. The four pieces of the knife are still in the possession of the Jung family.

Chapter 12
Creating Universes and Educational Environment

According to the contemporary science paradigm, the origins and the development of our Universe, life, and intelligences are the result of entirely random processes. The evolution of space civilizations, the Universe, and life depends on an improbable combination of astrophysical, geological, chemical, and biological events and circumstances.

Life and intelligence are not accidental by-products of our developing Universe.

The number and the level of development of space civilizations depend not on an improbable combination of various events and circumstances, but on the ideas and the works of a superior intelligence.

Who created the natural laws, the Universe, the Solar System, the Earth, and the intelligent beings? Are the physical laws that favor life and intelligence in our Universe the result of some sort of cosmic natural selection, an entirely random process, or were they programmed by a superior civilization? Many researchers suggest that life and intelligence are not some fancy cosmic accidents but they were programmed to happen.

The human DNA is strongly influenced by natural selection but now researchers are making tremendous efforts to reprogram it, even to create perfect artificial DNA in order to have perfect babies. In the near future human evolution will be entirely in the hands of man. In the far future the intelligent creatures will be able to guide the evolution of universes.

The fine-tuned Universe is the idea that our world is remarkably well suited for life and intelligence, to a degree that is unlikely to happen by mere chance. If some of the fundamental physical constants were to vary only slightly, the establishment and development of matter, astronomical structures, life, and intelligence as we know them would not have happened.

Alan Guth, John Gribbin, Andrei Linde, and many other researchers believe humans will in time also be able to create new universes. The physicists of the future will be the new intelligent designers.

Alan Guth, a theoretical physicist and cosmologist, author of inflation theory, said in the article "Physicist aims to create a universe, literally" by Malcolm W. Browne, published in The New York Times, 1987, "Such an achievement is obviously far beyond our technology but some advanced civilization in the distant future might. . . well, you never know. For all we know, our own universe may have started in someone's basement."

In 2002, Alan Guth said to the BBC, "I in fact have worked with several other people for some period of time on the question of whether or not it's in principle possible to create a new universe in the laboratory. Whether or not it really works we don't know for sure. It looks like it probably would work. It's actually safe to create a universe in your basement."

In 1991, the cosmologist and one of the main authors of the inflationary universe theory, Andrei Linde at Stanford University, published the article "Hard art of the universe creation." He suggested the possibility of creating a universe in a laboratory. Linde wrote, "Is it the reason why we must work so hard to understand strange features of our beautiful and imperfect world? Does this mean that our universe was created not by a divine design but by a physicist hacker? If it is true, then our results indicate that he did a very difficult job. Hopefully, he did not make too many mistakes."

Ray Kurzweil said with a pinch of humor, "Maybe our whole Universe is a science experiment of some junior high school student in another Universe. Given how things are going, she may not get a good grade."

The creators of our Universe were closer to men than to gods, argues John Gribbin. In his article "Are we living in a designer Universe?" published in *The Telegraph*, 2010, he said, "The argument over whether the Universe has a creator, and who that might be, is among the oldest in human history. But amid the raging arguments between believers and sceptics, one possibility has been almost ignored – the idea that the Universe around us was created by people very much like ourselves, using devices not too dissimilar to those available to scientists today."

When a civilization becomes advanced enough, it in turn creates its own universe. If humanity survives and develops enough, it also will create universes, life, and intelligences, and the new sentient creatures would wonder why their universe is fine-tuned. And why we don't visit them in an open way or officially?

Life on Earth has a very long evolutionary history, long before the origin of our home Universe. That's why life on our planet seems so incredibly successful and far beyond what evolution governed by chance events might offer.

St. Augustine wrote, "Miracles happen, not in opposition to Nature, but in opposition to what we know of Nature."

In many cases UFOs and paranormal objects and entities (angels, the devil, aliens, ghosts, demons, mythological creatures, etc.) are computer generated 3-D objects inserted into our material reality, which reflects the simulation. Sometimes the System defies the natural laws and UFOs, aliens, cars, horsemen, supernatural beasts, even common things materialize and dematerialize; UFOs are making impossible maneuvers; people levitate; tables, chairs, and other furniture move about by themselves; stones fall from the ceilings of rooms, etc. The laws of physics are just rules (part of numerous algorithms) for the simulation that in some cases can be changed allowing paranormal phenomena, religious, and other miracles.

Arthur C. Clarke said, "Magic's just science that we don't understand yet."

The paranormal manifestations are a result of mind control and insertion of materialized computer-generated objects or a combination of both. In some cases the UFOs leave physical traces behind: deep impressions found at the sites of landings, burned circles, etc. In a material world driven by a simulation, even UFOs, a result of mind manipulation, can leave physical traces.

The offspring of animals and humans are developing while playing. The paranormal phenomena are the games provided by the System for the evolving adult human animals in order to stimulate our thinking and the advance of science.

In the *Invisible College* Jacques Vallee wrote about where UFO and paranormal information comes from:

"A perceptive reader of [Andrija] Puharich's book [*Uri: A Journal of the Mystery of Uri Geller*] will note that the voice of SPECTRA consistently uses concepts that are current in the thoughts of either Uri or the author himself. In particular, it gets confused about astronomical units of measurement in precisely the same manner as Uri Geller does:

"Millions of light-years backwards into the dieshold [?] of the ages," says SPECTRA (p. 185), confusing units of time with units of space. And on another occasion the Rhombus 4-D computer states: "That was planned hundreds, hundreds of light-years ago, Andrija." And SPECTRA also spoke of "many billions of light-years ahead of time." Therefore we should consider the possibility that we are dealing with a phenomenon that uses, or emanates from, the brains of Geller and Puharich."

The phenomenon (actually the System) often negates itself in many ways to provoke our thinking. It is educating us, helping us to create genuine science. SPECTRA, one of the many disguises of the System, perfectly knows what a light-year is.

The System is fabricating a creative environment (scientific, cultural, religious, educational, emotional, mythological, etc.) in order to stimulate human intellectual development just as we are creating an educational environment with toys, fairy tales, books, pictures, Santa Claus, etc., in order to help our children develop as fast as possible and become more intelligent than us, their parents.

An educational toy should educate and stimulate mind growth. Learning and development are associated with interacting with the toy.

Our hidden teacher provides us with educational toys, too. In previous times instead of ufonauts, there were hordes of all sorts of fantastic creatures like demons, angels, devils, deities, mythological beasts, succubi and incubi, fairies, sirens, elves, gnomes, etc. The human mind has been intensively manipulated since the dawn of humanity. The UFO and paranormal phenomena are part of the controlled environment, which should stimulate human evolution and mind growth.

For the higher intelligence that is controlling us, humanity is still in its infant years. We are still part of the animal kingdom, animals with primitive computers.

Biblical wisdom says: "The fate of man and the fate of the animals is one and the same. As one dies, so does the other; the same breath is in them all. Man is no better than the animals. Both are bound for the same end; both sprang from the dust, and to the dust they both return," Ecclesiastes.

The immortal human soul is part of the tricks of the System, creating a psychologically more comfortable environment for man. We are doing the same for our kids. Santa Claus is part of a kids' "religion" created and supported by us. Actually, the Santa Claus lore was created by the System.

Humans are still not ready to accept the ugly truth about being a human.

The supernatural is just science that we don't understand yet.

Some individuals standing on the roof of the paranormal are jumping up and down shouting: "Materialism is dead! Materialism is dead!"

Various researchers are writing that materialism is dead because it cannot explain many supernatural phenomena. The problem is not in materialism but in our insufficient knowledge about reality. The problem is our still too primitive science—as the people so the science. Materialism has a long way to go; it is still in its infant years. There is no such thing as postmaterialist consciousness research or postmaterialist science, or spiritual science.

The spiritual is only in our minds. Outside of our minds everything is matter and energy (reflecting a simulation operated by the System) which we still don't understand well enough.

If we accept the System as part of our existence and the Universe, the paranormal becomes normal and part of materialism. It becomes subject of science, not of parapsychology. The future science will discover the physics behind the paranormal and the paranormal will become normal, and will become part of materialism.

Don't be misled by the fact that paranormal researchers are interpreting many facts wrongly and are producing a lot of nonsense writings and teachings. Science is producing a lot of nonsense, too. But this is the way that the quest for knowledge works.

Some scientists want all inexplicable and supernatural phenomena to be repeated in their laboratories, but science, scientists, and laboratories are still too primitive to test every aspect of the Universe, life, and intelligence. This will be done by future science, scientists, and laboratories.

Skepticism in science is healthy and it is a must, but it requires due intelligence and enough knowledge because when they are lacking, skepticism turns into paranoia and pseudoscience.

If you want to believe, enter some teaching or doctrine—religious, spiritual, pseudoscientific, New Age, UFO cult, etc. If you want to explore the world, get rid of all teachings, doctrines, and beliefs, all of them. They could be only objects of your research. And don't forget a phrase from Ecclesiastes: "in much wisdom is much grief, and he who increases knowledge increases sorrow."

To some people, a Universe controlled by a simulation and a System seems living, conscious, and intelligent. A number of scholars advocate the idea that the Universe is governed by a supernatural being, a supreme spiritual force, or that it is even a part of the mind of some god.

The educational environment for adults is multifaceted and the paranormal phenomena, UFOs, religions, cults, etc. are part of it. One can't create a realistic picture of our world by dismissing them.

Sometimes people wonder why the aliens attempting to conceal alleged encounters with extraterrestrials seem so inefficient at blotting out human memory. Even amateur hypnotists can unblock it.

Many researchers use hypnosis to reveal UFO and alien abduction events, expecting to get to the genuine nature of what really happened. But one should know that actually they are revealing the controlled unconscious of the person—memories are implanted; pictures, thoughts, and emotions are superimposed on real events. Even during the hypnosis sessions the minds of the hypnotist and the hypnotee are controlled and manipulated by the System.

Statistics reveal that about half of alien abduction cases were discovered under regressive hypnosis. The most impressive UFO encounters and abduction cases were reported by people under hypnosis. The alien abduction and UFO "memories" were inserted by the System into the unconscious mind of the "witnesses" and they were revived by the hypnotists. In some cases the abduction "memories" were actually created by the hypnotists from visual material memorized by the "abductee" from movies, TV, books, newspaper pictures and articles, etc. Some hypnotists actually plant suggestions into the mind of the abductees and contactees.

Witnesses of UFOs, alien abductions, and paranormal manifestation successfully pass lie-detector tests. The control of their minds by the agency is so successful that they cannot tell what is real and what is manipulation of their minds.

In a world controlled by the System and a simulation there is no such thing as reliable witness. The minds of all people on Earth are controllable and are controlled. They will see, hear, feel, imagine, and think whatever the System decides.

If you think that manipulating the human mind is something very sophisticated and difficult, let's see what an ordinary stage magician from 19th century can do.

The Scottish surgeon James Braid described in 1851 the hypnotic stage performance by a visiting American, George W. Stone:

"Persons in a perfectly wakeful state, of well-known character and standing in society, who come forward voluntarily from among the audience, will be experimented upon. They will be deprived of the power of speech, hearing, sight. Their voluntary motions will be completely controlled, so that, they can neither rise up nor sit down, except at the will of the operator; their memory will be taken away, so that they will forget their own name and that of their most intimate friends; they will be made to stammer, and to feel pain in any part of their body at the option of the operator – a walking stick will be made to appear a snake, the taste of water will be changed to vinegar, honey, coffee, milk, brandy, wormwood, lemonade, etc., etc., etc. These extraordinary experiments are really and truly performed without the aid of trick, collusion, or deception, in the slightest possible degree."

A regular stage magician can make you see and meet the entire menagerie of UFOs, close encounters of the third kind, alien abductions, deities and demons, mythological creatures, etc.

Both believers and skeptics are being manipulated. The System creates supporters and opposition of everything—sports clubs, scientific theories, religions, states, armies, criminal gangs, etc. They are part of the game called competition which is part of the larger game called fast evolution. Someone is expecting a result from our developing Universe. What the end result would be, we don't know. Possibly a superior intelligence.

In many cases the System creates UFO events with religious and paranormal elements, to the great disappointment of the buffs who believe that the UFOs are alien spacecraft.

Science still does not have hard evidence of real extraterrestrial visitations, but this does not rule them out. However, such a possibility has nothing to do with the UFO phenomenon.

Glossolalia, referred also as "speaking in tongues," has been around for thousands of years. This is a phenomenon in which people speak languages that are unknown to the speaker.

Andrew Newberg, MD, Associate Professor of Radiology, Psychiatry, and

Religious Studies, wrote, "Our finding of decreased activity in the frontal lobes during the practice of speaking in tongues is fascinating because these subjects truly believe that the spirit of God is moving through them and controlling them to speak. Our brain imaging research shows us that these subjects are not in control of the usual language centers during this activity, which is consistent with their description of a lack of intentional control while speaking in tongues."

"These findings could be interpreted as the subject's sense of self being taken over by something else."

Entities like aliens, angels, spirits, demons, etc. are able to quote the Bible in Latin, Hebrew, Greek, and in many other tongues. Actually, they know all languages on Earth, living or long dead. These entities seem to have a remarkable knowledge of people's personal business. The devil and the spirits in religious cases of possession know every single detail about the lives of their exorcists. They know everything—past, present, and future. Actually, the one who knows everything is the System. The aliens, angels, demons, spirits, etc. are its creations. The System empowers them in many ways, like creating miracles, lore, myths, bodies of followers and disciples, etc.

Many mediums are simple, uneducated people, but during séances they are able to speak foreign languages or play piano.

The thirteen-year-old Laura, daughter of Judge Edmonds, president of the Senate in the 1850s, reportedly played sophisticated Beethoven pieces while in a trance, even though she had never previously touched a piano. Mediums who knew only English during séances spoke Greek, Italian, Arabic, Chinese, German, and other languages fluently.

Sometimes the aliens and the spirits look at us as silly people and their behavior is utterly moronic, but this is just a game of the agency which is tremendously intelligent and possesses knowledge far beyond our own. The paranormal entities and UFO aliens were playing their weird games because they are puppets on strings.

People who have contacted such entities know that the agency also is gifted with a mischievous sense of humor and sometimes plays practical jokes.

The System also fabricates all sorts of speculations and conspiracy theories (like ancient astronauts, flat Earth, hollow Earth, suppression by the authorities of perpetual motion and cold fusion technology, many political leaders are aliens, the Philadelphia Teleportation Experiment, the US Moon landings were staged by NASA in a film studio, etc.) in order to stimulate rational thinking and science by creating and supporting competitive (often fake) theories, opinions, positions, groups, all sort of proponents and skeptics, etc.

The System supports religions by creating miracles. One such miracle was the Zeitoun apparitions on the roof of St. Mary's Coptic Church in the Zeitoun district of Cairo, Egypt.

In 1968, two bus mechanics saw something strange on the roof of the Coptic Church. They thought someone was going to commit suicide, but after closer examination, it was a silhouette of light that resembled a woman dressed in white. The apparition lasted a few minutes.

One week later, the phenomenon reoccurred again lasting for a few minutes. After that, apparitions became more frequent, sometimes two or three times a week, for several years, ending in 1971.

This Marian apparition in Egypt was witnessed by at least several hundred thousand people.

The head of the Coptic Church investigated the apparitions, and he declared them miraculous and an authentic appearance of the Virgin Mary. A Vatican envoy arrived in Cairo and made their own report. The Papal Residence in Cairo made an official statement:

"Since the evening of Tuesday April 2, 1968, the apparitions of the Holy Virgin Saint Mary, Mother of Light, have continued in the Coptic Orthodox Church named after Her in Zeitoun, Cairo.

The apparitions occurred on many different nights and are continuing in different forms. The Holy Virgin Saint Mary appeared sometimes in full form and sometimes in a bust, surrounded with a halo of shining light. She was seen at times on the openings of the domes on the roof of the church, and at other times outside the domes, moving and walking on the roof of the church and over the domes. When She knelt in reverence in front of the cross, the cross shone with bright light. Waving Her blessed hands and nodding Her holy head, She blessed the people who gathered to observe the miracle. She appeared sometimes in the form of a body like a very bright cloud, and sometimes as a figure of light preceded with heavenly bodies shaped like doves moving at high speeds. The apparitions continued for long periods, up to 2 hours and 15 minutes...

Thousands of people from different denominations and religions, Egyptians and foreign visitors, clergy and scientists, from different classes and professions, all observed the apparitions. The description of each apparition as of the time, location and configuration was identically witnessed by all people, which makes this apparition unique and sublime.

Two important aspects accompanied these apparitions: The first is an incredible revival of the faith in God, the other world and the saints, leading to repentance and conversion of many who strayed away from the faith. The second are the numerous miracles of healing which were verified by many physicians to be miraculous in nature.

The Papal Residence has thoroughly investigated the apparitions and gathered information by way of committees of clergy who have also witnessed the apparitions by themselves and recorded everything in reports presented to His Holiness Pope Kyrillos VI.

...May God make this miracle a symbol of peace for the world, and a blessing for our nation as it has been prophesized: "Blessed be Egypt My people."

Saturday May 4, 1968
Papal Residence in Cairo"

The apparitions were captured by newspaper photographers and Egyptian television. Investigations performed by the police could find no apparent explanation. No device was found within a radius of fifteen miles capable of projecting the image. With no alternative explanation and approval from religious and political leaders, the Egyptian government accepted the apparitions as a true religious miracle.

Many photos were taken of the apparition from independent sources.

Like some UFOs, a number of religious miracles are materializations by the System and can be photographed— unlike the Fatima case which was a result of a mind control.

On the other hand, such physical materializations can be interpreted differently by some witnesses because of the mind control of the System. Some can see only a blurry figure of light, others can "see" clearly the Virgin Mary.

The System keeps some issues ambiguous—open to or having several possible interpretations to stimulate human thinking and science.

The System creates phenomena and at the same time it denies them.

Some people get useful advice even from ufonauts. Pat McGuire, a farmer, many times observed UFOs hovering over his land. One night aliens took him aboard their spaceship. He recalled the abduction under hypnosis. The aliens told him to drill a well in high plains country near his ranch. McGuire consulted specialists, but drilling experts and geologists said that there was no water there because the place was too high. The neighbors called him crazy. But McGuire bought the land in the high plains near his ranch, drilled a well, and got plenty of water.

Of course, instead of advice from a ufonaut, the farmer could have had a dream in which angels were successfully drilling for a well in the high plains country near his ranch, or his deceased uncle could have come as a ghost to give him instructions on where to find water. The information from the System can be conveyed in many picturesque and highly impressive ways.

Mikhail Lomonosov was a polymath (a person of great and varied learning), scientist, and writer, who made important contributions to literature, education, and science. Among his discoveries was the atmosphere of Venus.

In 1741, returning from Germany after graduation, Lomonosov had a dream. He saw his father, Vasily, in a terrible shipwreck. Vasily died on an obscure island. In his dream, the father asked his son to give him a proper burial.

When back home, Lomonosov visited some local fishermen and told them the exact place where to look for his missing father. They went to the island, found Vasily's dead body, and buried him.

But one should be very careful when using such information, because alongside useful health, professional, or business advice, one inevitably would receive a mountain of trash, especially if he/she asks too many questions. For thousands of years experienced occultists have known that man should invoke young, inexperienced spirits because the old ones cheat too much. Some people, encouraged by channeled information that helped them to improve health or business, begin to believe everything they receive from the mystery source, and sometimes the results are disastrous. They lose jobs, get divorced, become laughing stocks, lose money, even die.

The Swedish scientist, engineer, and mystic Emanuel Swedenborg warned, "When

Spirits begin to speak with a man, he must beware that he believe nothing that they say. For nearly everything they say is fabricated by them, and they lie..."

Alien visitors say repeatedly, "You should believe in us but not too much."

Uri Geller also had his doubts about the voice that told him what to do and what not to do, and began wondering if it might be just "a goddam little clown that is playing with us."

Croesus, the king of Lydia, was renowned for his wealth. According to Herodotus, Croesus desired to discover who of the well-known oracles was most trustworthy. He sent messengers to the oracles ordering that on the 100th day from their departure they should ask what Croesus was doing on this exact date. Only the oracle at Delphi answered correctly, telling the messenger that his king was boiling a tortoise and a lamb in a cauldron.

Croesus began preparing a war against Persia. Before setting out, he turned to the Delphic oracle to inquire whether he should invade Persia. The oracle replied that following the war a great empire would be destroyed. After such a prediction, Croesus was sure that he would win the war and launched the invasion. However, Persia was victorious and the doomed empire was Lydia. The powerful empire destroyed by the war was Croesus's own.

Never trust channeled information. You will never know whether is it correct or wrong until the events happen.

Even the best oracles never know the truth. They just repeat what the System is telling them.

One more thing—your enemies also receive channeled information and predictions of future events; they also ask the divine power to win the battle and you to be destroyed. The System does not work for some of the states, kings, leaders, or persons—it works for itself and its master, the superior intelligence. Persons, states, cultures are just fleeting moments in the long history of the developing intelligence.

Demons, alien visitors, saints, gods and goddesses, voices and telepathically transmitted thoughts from mysterious sources will promise you spiritual freedom, ultimate knowledge, and immortality, but instead becoming rich, free, wise, and immortal, you will become a slave of the entity. Some contactees, mediums, UFO witnesses, and prophets ultimately end up sitting on hilltops to await the end of the world and the flying saucer guys or the god that will rescue them. But they never come!

Obviously, the mysterious source of information is not only a giant database, but there is an intelligence behind it and it can mock people and make practical jokes. This intelligent entity likes ridiculing humans. This was confirmed by many contactees and paranormal witnesses.

Most things on Earth seem wrong and foolish from a human point of view, and many people are convinced that there are no superior creatures controlling life and intelligence on our planet. Yes, things are wrong and foolish from our point of view, but from the point of view of the controlling superior intelligence they are right because the things we accept as wrong and foolish are actually signs for high-level competition and harsh education, exactly what is expected by the masters. Otherwise we can conclude that life on Earth is some sort of sinister computer game organized and owned by some freak extraterrestrial superior civilization or individual enjoying the misery of human beings.

Fast development of the sentient creatures of our Universe is an imperative.

The information from the System comes to people in different ways: dreams, visions, sensing thoughts, lucid dreaming, intuition, automatic writing, hearing voices, using a pendulum, dowsing, etc. People can't really control the channel of information. It is so constructed that in most cases men can't tell where the information comes from—from their mind or from an external agency.

There are millions of cases proving that telepathy is real. Then why can't we use it at will in our everyday life? Because our world is based on hard competition. If there is telepathy for everyday use, then there will be no real competition because there will be no secrets, the basis of competition, crime, wars, etc.

Chapter 13
Precognition of the Future and Determinism

The past exists somewhere as a detailed record. Sometimes people can see this recorded past.

A *time slip* is a phenomenon in which a person, or group of people, seem to travel back through time and can see the past version of a given environment.

One of the best-known, and earliest, examples of a time slip was reported by two Englishwomen, Charlotte Moberly and Eleanor Jourdain, who claimed that they slipped back in time in the gardens of the Petit Trianon at Versailles from the summer of 1901 to the period of the French Revolution (from 1789 until 1799).

They travelled with a guidebook, but the two women soon became lost. They entered a lane, where Moberly noticed a woman shaking a white cloth out of a window and Jourdain noticed an old deserted farmhouse, outside of which was an old plough. At this point they claimed that a feeling of oppression and dreariness came over them. They then saw some men who looked like palace gardeners, who told them to go straight on. Moberly later described the men as "very dignified officials, dressed in long greyish-green coats with small three-cornered hats." Jourdain noticed a cottage with a woman and a girl in the doorway. The woman was holding out a jug to the girl. Jourdain described it as a "tableau vivant," a living picture, much like Madame Tussaud's waxworks. Moberly did not observe the cottage, but felt the atmosphere change. She wrote: "Everything suddenly looked unnatural, therefore unpleasant; even the trees seemed to become flat and lifeless, like wood worked in tapestry. There were no effects of light and shade, and no wind stirred the trees."

They reached the gardens in front of the palace. Moberly noticed a lady sketching on the grass who looked at them. She later described what she saw in great detail: the lady was wearing a light summer dress, on her head was a shady white hat, and she had lots of fair hair. Moberly thought she was a tourist at first, but the dress appeared to be old-fashioned. Moberly came to believe that the lady was Marie Antoinette. Jourdain, however, did not see the lady.

Moberly did not observe the cottage that Jourdain had noticed. Moberly saw Marie

Antoinette; Jourdain, however, did not see the lady.

That was no real time travel and the past is not still present in another dimension or parallel universe as some researchers claim; the past the both women "visited" was an individually interpreted vision in a dreamlike state of the recorded past.

So, what about the future?

Probably the System is rendering a simulation of the future of the Universe, life, and intelligence to provide a fast advance of individuals and civilizations, and to prevent self-annihilation when the intelligent creatures of the Universe invent weapons and technologies able to delete entire planets and space races. In the future, there will be numerous wars with super-weapons.

Precognitions come to us in the form of dreams, literary fiction, voices or thoughts in the head, fortune-telling, visions, etc.

A common mistake of some contemporary scientists is that precognition violates the principle of causality, an effect cannot occur before its cause.

Many researchers claim that past, present, and future exist simultaneously in material form.

The past and the future (the simulation) are only records, saved information. They are not material. Only the present is material.

The past is settled and unchangeable, but the digital model of the future is unsettled and dynamic. The many-worlds interpretation should imply that potential alternate histories are possible digital models of the future which will result in a single material present.

Since the future is not material and it didn't happen yet, precognition does not violate the principle of causality; an effect does not occur before its cause.

The future is predictable, but only in cases when the System decides to give us information about some events to come. Humans don't have direct access to the simulation of the future.

Some phenomena are influenced by supposed time travel. A number of UFO contactees were given a time travel tour into the past, including meeting with prominent figures of that time like Jesus.

The time-slip phenomenon is also connected with time travel—people are involuntarily transported back in time.

The past and the future are records and sometimes we are allowed to see in visions, dreams, lucid dreams, during medium séances, etc. past events or events to come.

Such manifestations are part of the never-ending manipulation of the human mind by the System.

We can't really travel backward or forward in time.

If we have direct access to the simulation of the future and the record of the past, we can time travel virtually. With the appropriate technology some superior intelligence could materialize a given virtual time period of the past and the future in order to visit them in reality.

John Burrows wrote in his book *Afterlife - the Proof* about psychometry.

"[Stefan] Ossowiecki always worked the same. He would take the object in his hands and concentrate until the room before him, and even his own body, became shadowy and almost nonexistent. After this transition occurred, he would find himself looking at a three-dimensional movie of the past. He could then go anywhere he wanted in the scene and see anything he chose. While he was gazing into the past, Ossowiecki even moved his eyes back and forth as if the things he was describing possessed an actual physical presence before him."

There is a lot of evidence that there is a record of the past which some psychics can read (actually receive information from the System where everything is recorded).

Maybe somewhere in the distant future humans could have unlimited access to the record of the past or even of the future (hardly)! Or maybe in the far future the humans will have free will and the future of man will be not predestined. We cannot know.

In the Fatima case, there were also precognitions. Among the prophecies was the prediction that Francisco and Jacinta would die soon. But the kids were not bothered. On the contrary, they were delighted because it meant that they would be going to heaven. In the book *Our Lady of Fatima,* William T. Walsh wrote about the Francisco's attitude to the prediction.

"Do you want to be a carpenter?'

"No, ma'am."

"Do you want to be a soldier, then?"

"No, ma'am."

"A doctor, isn't that it?"

"Oh, no!"

"I know what you would like be—a priest!"

"No."

"What! To say Mass?... to hear confessions?...to pray in the church? Isn't that it?"

"No, *senhora*. I don't want to be a *padre*."

"Then what do you want to be?"

"I don't want to be anything."

"You don't want to be anything?!"

"No. I want to die and go to heaven."

The prediction was fulfilled in time and both children died.

Johann Wolfgang von Goethe, in his autobiography *Dichtung und Wahrheit ("Poetry and Truth")*, wrote how, having just said goodbye to his girlfriend and riding along a road, when he saw his doppelgänger—a look-alike or double of a living person, sometimes portrayed as a paranormal phenomenon, and in some traditions as a harbinger of bad luck.

Goethe wrote, "Amid all this pressure and confusion I could not forego seeing Frederica once more. Those were painful days, the memory of which has not remained with me. When I reached her my hand from my horse, the tears stood in her eyes; and I felt very uneasy. I now rode along the foot-path toward Drusenheim, and here one of the most singular forebodings took possession of me. I saw, not with the eyes of the body, but with those of the mind, my own figure coming toward me, on horseback, and on the same road, attired in a dress which I had never worn—it was pike-gray, with somewhat of gold. As soon as I shook myself out of this dream, the figure had entirely disappeared. It is strange, however, that, eight years afterward, I found myself on the very road, to pay one more visit to Frederica, in the dress of which I had dreamed, and which I wore, not from choice, but by accident."

Goethe had seen himself eight years in the future.

Michel de Nostredame (1503–1566), popular as Nostradamus, was a French apothecary and reputed seer who published collections of prophecies that have since become widely famous.

Here is one of his most popular precognitions.

"The young lion will overcome the older one,

On the field of combat in a single battle;

He will pierce his eyes through a golden cage,

Two wounds made one, then he dies a cruel death."

King Henry II of France was an avid hunter and a participant in jousts and tournaments.

Henry II and the Comte de Montgomery, captain of his elite Scottish Guard, had lions on their shields. On 30 June 1559, Henry II (the older one) lined up to joust Montgomery (the young lion). In their final pass, Montgomery's lance tilted up, and burst through the king's gilded visor (golden cage) splintering to pieces. Two shards, one through the eye and one through the temple, lodged in the king's head.

Despite the efforts of the royal surgeon, the king died of septicemia (infection of the blood) on 10 July 1559.

One night in 1858 at his sister's home in St. Louis, Mark Twain had a precognitive dream of the death of his brother Henry. Both were laboring aboard the steamboat *Pennsylvania* which sailed the Mississippi river.

Mark Twain wrote in his book *Autobiography of Mark Twain*, Volume 1, "In the morning, when I awoke, I had been dreaming, and the dream was so vivid, so like reality, that it deceived me, and I thought it was real. In the dream I had seen Henry a corpse. He lay in a metallic burial case. He was dressed in a suit of my clothing, and on his breast lay a great bouquet of flowers, mainly white roses, with a red rose in the center. The casket stood upon a couple of chairs."

Two weeks later, Mark Twain was transferred to another riverboat while Henry remained aboard the *Pennsylvania*.

Shortly thereafter, the four boilers of the *Pennsylvania* violently exploded and most of the crew members were critically wounded, including Henry, who died a few days later.

When Mark Twain entered the funeral parlor, he was horrified to see the picture of his dream: his dead brother laid out in a metal casket in a borrowed suit. The casket stood upon a couple of chairs. Only the bouquet was missing. As he watched and mourned, a nice young lady came in with a bouquet of white roses with a single red one at the center and laid it on Henry's chest.

The well-documented Fatima Miracle of the Sun was predicted by a female apparition several months upfront.

The next case is from *Assigned to Adventure*, 1938, by Irene Kuhn.

"Mrs. Kuhn, an American journalist, had married a fellow American reporter when both were working in China. Mrs. Kuhn returned to

America for a holiday, leaving her husband in China. One December afternoon, while walking on Michigan Boulevard in Chicago, the ordinary physical surroundings suddenly vanished and she experienced a "vision." She saw "a strip of green grass within a fence of iron palings. Three young trees, in spring verdure, stood at one side; beyond the trees and the fence, in the far distance, factory smokestacks trailed sooty plumes across the sky." She saw a small group of men and women dressed in black clothes; a limousine drew up on a graveled road by the grass. Two men alighted and offered their hands to a woman in black. The woman was herself. She was gently urged by the men towards the group and she then saw a small hole in the grass into which someone was placing a small box. She then recognized the group of people as members of her husband's family. Only the family were there—her husband was missing. Then she knew what was in the box and she "crumpled on die grass without a sound.""

The precognitive vision was fulfilled precisely about five months later, in the early spring.

Spencer Perceval, the British Prime Minister, was killed in the lobby of the House of

Commons in London on 11 May 1812 by John Bellingham, a merchant with a grievance against the government.

The Times reported that a Cornish industrialist, John Williams, received a dream warning of Perceval's assassination on 2 or 3 May 1812, nearly ten days before the event.

John Williams wrote and signed the following statement, "My dream was as follows: About the second or third day of May, 1812, I dreamed that I was in the lobby of the House of Commons (a place well known to me). A small man, dressed in a blue coat and a white waistcoat, entered, and immediately I saw a person whom I had observed on my first entrance, dressed in a snuff-coloured coat with metal buttons, take a pistol from under his coat and present it at the little man above-mentioned. The pistol was discharged, and the ball entered under the left breast of the person at whom it was directed. I saw the blood issue from the place where the ball had struck him, his countenance instantly altered, and he fell to the ground. Upon inquiry who the sufferer might be, I was informed that he was the chancellor. I understood him to be Mr. Perceval, who was [had formerly been] Chancellor of the Exchequer. I further saw the murderer laid hold of by several of the gentlemen in the room.

Upon waking I told the particulars above related to my wife; she treated the matter lightly, and desired me to go to sleep, saying it was only a dream. I soon fell asleep again, and again the dream presented itself with precisely the same circumstances. After waking a second time and stating the matter again to my wife, she only repeated her request that I would compose myself and dismiss the subject from my mind. Upon my falling asleep the third time, the same dream without any alteration was repeated, and I awoke, as on the former occasions, in great agitation. So much alarmed and impressed was I with the circumstances above related, that I felt much doubt whether it was not my duty to take a journey to London and communicate upon the subject with the party principally concerned. Upon this point I consulted with some friends whom I met on business at the Godolphin mine on the following day. After having stated to them the particulars of the dream itself and what were my own feelings in relation to it, they dissuaded me from my purpose, saying I might expose myself to contempt and vexation, or be taken up as a fanatic. Upon this I said no more, but anxiously watched the newspapers every evening as the post arrived.

"On the evening of the 13th of May (as far as I recollect) no account of Mr. Perceval's death was in the newspapers, but my second son, returning from Truro, came in a hurried manner into the room where I was sitting and exclaimed: 'O, father, your dream has come true! Mr. Perceval has been shot in the lobby of the House of Commons; there is an account come from London to Truro written after the newspapers were printed.'"

The British Prime Minister, Perceval, himself had a series of dreams culminating on 10 May with one of his own death, which he had while spending the night at the house of the Earl of Harrowby. He told the Earl of his dream, and the Earl tried to persuade Perceval not to attend Parliament that day, but Perceval refused to be scared off by "a mere dream" and headed for Westminster on the afternoon of 11 May.

He was shot dead in the lobby of the House of Commons at about 5:15 PM.

The source of the next case is the book *Foreknowledge* by H. F. Saltmarsh, 1938. The original text from *Some Cases of Prediction* by Alfred Lyttelton (Mrs. Edith Balfour Lyttelton), 1937.

"Mr. Calder, the Headmaster of the Grammar School at Goole, wrote to Mrs.

Lyttelton, after her broadcast talk on precognition, giving a full account of two precognitive dreams experienced by his wife.

In 1928 he was appointed headmaster of Holmforth Secondary School in Yorkshire. Before leaving Middlesex, where they then resided, Mrs. Calder, who had never been to Yorkshire, dreamed of an old greystone house, set in a lovely valley through which ran a stream of clear but black-looking water. In their house-hunting near Holmforth they came across the very house which Mrs. Calder had seen in her dream; they took it, or rather one half of it, and moved in in August 1928. They found that the water of the stream was frequently discolored by indigo from a near-by dye-works. In her dream Mrs. Calder had seen that only one half of the house was occupied and that outside the door of that half was a barrel which was used as a dog kennel. When they went to live there, though the other half was occupied, there was no barrel. A year or so later, there was a change of tenants of the other half of the house. When the new people arrived they brought with them a dog and placed a barrel outside the door for its kennel."

The next case is from a letter written by Mademoiselle Dudlay of the Comedie Francaise about the young actress Irene Muza who was burned to death in 1909.

Irene was a convinced spiritualist and several months before the incident during a séance, in which she was in a deep hypnotic sleep, they asked her if she saw what awaited her, personally, in the future.

She wrote the following, "My career will be short. I dare not say what my end will be. It will be terrible."

The experimenters, much impressed, erased the words before she awakened; thus, consciously at least, she never knew what a terrible thing she had predicted for herself.

Several months afterward her hairdresser was sprinkling her hair with an antiseptic lotion made of mineral essences, when she let several drops of the liquid fall on the lighted stove. These instantly flamed up; fire enveloped the hair and clothing of the actress, who in a second was wrapped in flame and suffered such burns that she died at the hospital a few hours later.

In 1981 British Rail had a call from a woman who claimed to have had a vision of a fatal crash in which a freight train had been involved. So clear had it been, she said, that she not merely saw the blue diesel engine, but could read the number: 47 216.

Two years later, an accident of the kind she predicted occurred, all the details matching—except one: the engine's number was 47 299.

That would have been that, but a train spotter, Howard Johnston, happened to have noticed that 47 299 was not the engine's original number. It had been renumbered, a couple of years before, from 47 216. Diesels, he knew, were ordinarily renumbered only after major modifications, which this one had not undergone. When curiosity prompted him to ask why, he was told about the prediction.

Apparently British Rail officials had been sufficiently impressed (they had checked with the local police, and found that the woman who had provided it had given them some useful information from her visions) to try to ward off Fate by changing the number. The ruse had failed, and they had officially logged it all as an "amazing coincidence."

The case is from the book *Coincidence: A Matter of Chance – or Synchronicity?* by Brian Inglis. According to the author, he has compiled a collection of fascinating accounts that will leave the most committed skeptics scratching their heads.

P.M.H. Atwater recounts the story of a woman with precognitive vision in her book *Future Memories*.

"I was doing the morning dishes when this rush of energy nearly lifted my head off. I suddenly experienced myself at a dinner party that night, saw who would be there, and took part in what happened and what was said. The whole thing was so real, I decided to make no plans for the evening, just to see what might happen. Sure enough, a friend called and began apologizing all over herself for being so tardy, then she asked if I would come to her dinner party that night. I had to muffle laughter as I accepted her invitation. When I arrived at the party, it was a duplicate of what I had already experienced that morning: every conversation, every wave of a hand, repeated what I previously lived through. I'm glad I 'attended' the dinner party before it happened so I could be prepared in advance."

The next two cases are from *Extraordinary Knowing: Science, Skepticism, and the Inexplicable Powers of the Human Mind* by Elizabeth Lloyd Mayer.

"I was on a bus and all of a sudden found myself smelling the perfume my brother's ex-wife used to wear. When the bus stopped, she got on. I hadn't smelled that perfume or seen her in thirty years."

"My husband and I fell in love with a house in London on our honeymoon—very distinctive, across from a park. Fourteen years later, living in Boston, I woke up one day, and thought, maybe we could buy it. I tracked down a realtor in London, asked if she could figure out the address and find out if it was for sale. Crazy! But she did. The man who'd been living there had just died; the For Sale sign wasn't even up yet. We bought it the next week."

The following case comes from the book *Closer to the Light: Learning from Children's Near-Death Experiences* by Melvin Morse and Paul Perry.

Jane had a pacemaker. The day before she died Jane was having a coffee at the kitchen table, when her dead sister appeared and said, "June, it's time to go." Then the apparition sat down at the table and drank a cup of coffee. When the apparition of her sister finished, she got up and left the house.

June felt that she couldn't tell her husband what had happened. Instead, she called the aunt and uncle who had raised her and told them about the visit of her dead sister.

June said that she is going to die, and just wanted to say goodbye.

Then she called her two brothers. She told them not to mention their conversation to Don until she died.

June told her husband how happy he had made her and that she was very glad to have such a beautiful home and a wonderful child. Nothing could have made her happier.

"That night she died in her sleep because her pacemaker failed. Her heart simply stopped."

Raymond Moody wrote about an interesting prediction during near-death experiences in his book *The Light Beyond.*

"The year was 1975, several months before the publication of *Life After Life*. It was Halloween and my wife at the time, Louise, had taken our children out trick-or-treating.

"They arrived at one house and were greeted by a pleasant man and woman who began to talk to the children. They asked the children their names and when my oldest said, "Raymond Avery Moody, the third," the woman looked startled.

"I must talk to your husband," she said to Louise.

"When I spoke to this woman later on, she told me about her near-death experience in 1971. She'd had heart failure and lung collapse during surgery and had been clinically dead for a long time. During this experience, she met a guide who took her through a life review and gave her information about the future. Toward the end of the experience, she was shown a picture of me, given my full name, and told that "when the time was right," she would tell me her story."

The small boy never said his full name: Raymond Avery Moody, the third. This was the first time he had ever given out the full name.

In 1988, September 2nd, Elizabeth Balkin (now Krohn) was struck by lightning in the parking lot of her synagogue in Houston. She had near-death experiences.

After that Elizabeth began having dreams about plane crashes or other disasters reported in the news. She authenticated her precognitions by recording them in emails sent to herself.

Elizabeth G. Krohn wrote about her experiences in her book *Changed in a Flash*, published in 2018. Her co-author was Jeffrey J. Kripal, a religions college professor. The following case is from the book.

On January 15, 2009, during a trip to Jerusalem, at 2:57 PM Israel time, she emailed herself the following dream after a brief afternoon nap:

"MID-SIZE COMMERCIAL PASSENGER JET (80-150 PEOPLE) CRASHES IN NYC. MAYBE IN RIVER. NOT CONTINENTAL AIRLINES. NOT AMERICAN AIRLINES.

IT IS AN AMERICAN CARRIER LIKE SOUTHWEST OR US AIRWAYS."

Her husband Matt did awake.

"OK. What did you see?" Matt asked.

"It's really weird," I said. "I saw this plane and it was sitting, kind of floating, on water, and there were people standing on the wings of the plane."

"This may be one that you miss," he said. "Airplanes are heavy. They 'float' like a rock. And people standing on the wings? … You saw people on the wings of a floating airplane?"

"Yes," I said.

"The physics of that are impossible," Matt said.

Six and a half hours later, US Airways Flight 1549, made an emergency landing in the Hudson River after it struck a flock of geese and lost engine power.

All 155 passengers survived. Photos that appeared in the news around the world showed the plane floating in the water; many passengers were standing on both wings.

In his book *Physics of the Impossible: A Scientific Exploration into the World of Phasers, Force Fields, Teleportation, and Time Travel* Michio Kaku wrote, "In summary, precognition is ruled out by Newtonian physics... It would set off a major shake-up in the very foundations of modern physics if precognition was ever proved in reproducible experiments."

According to most scientists, precognition goes against established principles of science of the present day. They deny the very existence of precognition.

I have a precognition story of my own experience.

A psychic woman and I were sitting at a café, having a chat. Suddenly, she told me that there was a problem with my son's car—there was a small fissure in the axle of the left rear wheel and the wheel would separate from the vehicle. Soon after my son sold the car because he bought a new one. After a few weeks he got a phone call from the owner of his previous car who told him that the left rear wheel separated from the vehicle.

I think that it was demonstrated to me that precognitions do indeed exist.

What should I believe in: physicists' theories, rejecting precognition, or my own experience? And the experience of millions of people around the world for thousands of years.

Is it possible that both sides are right? That precognition exists and it is not breaking the laws of nature? At first glance, this seems impossible. Most scientists are adamant that precognition is ruled out by contemporary physics.

Often precognitions are of such detail and specificity that accident or coincidence seemed to be ruled out.

Present science still can't explain precognition, telepathy, events like the Fatima case, etc. There are millions of cases of such phenomena experienced by ordinary people and psychics. Most scientists are still afraid to consider such cases. They behave like a spoiled child and just deny their existence.

"I shall not commit the fashionable stupidity of regarding everything I cannot explain as a fraud," wrote Carl Jung.

According to Wikipedia, precognition (from the Latin *præ-*, "before" and *cognitio*, "acquiring knowledge"), also called prescience, future vision, future sight, is an alleged psychic ability to see events in the future. But numerous examples show that many ordinary people with no psychic ability had spontaneous precognitive visions, dreams, or they hear warning voices in their heads about impending dangers.

The Fatima case shows that 100,000 ordinary people had visions of supernatural events brought to them by telepathic means. Telepathy is a communication of thoughts, images, sounds, smells, or ideas by means other than the known senses. From the Fatima case we know that telepathy is real and all people (with a few exceptions) can experience it. The Fatima events were *predicted* several months upfront by a small shepherd girl.

Precognition is most difficult to accept. Telepathy, psychokinesis, near-death experiences, even reincarnation can be explained in some way or other, but not precognition.

Physics denies it. The natural laws of present-day science do not suppose the existence of precognition.

Some researchers are looking for natural laws that would allow precognition—for instance, particles/waves that for some reason travel backward in time, and these hypothetical particles/waves carry information about the future. The observational theory proposes that the human brain can influence probabilistic events, determining some future events that way; the observer is actually "causing" these events, making them much more probable and they happen more often. A self-fulfilling prophecies theory suggests that the precognitive experience (dreams, visions, etc.) itself unleashes a powerful psychokinetic energy, which then brings the envisioned future to pass. Another theory suggests that the future already exists in a material form and the brains of people from future and present are somehow linked, and sometimes through this link we receive information about events yet to happen.

But we still haven't discovered natural laws that would allow the existence of precognition.

Maybe the point is not in the physics, but in the existence of a superior civilization that simulates the future. Our material world follows the simulation.

Riddles like the fine-tuned Universe, the Fermi paradox, the Fatima case, knowing future events, encounters with nonexistent UFOs, ufonauts, religious, and mythological creatures, etc. can be resolved successfully only accepting the existence of a creative superior intelligence and simulation of our material world. If we approach the precognition conundrum the same way, it is solvable.

Multiple teams of scientists claim that their version of the history of the Universe is right. But we should never forget that throughout the history of the All were created a great many extremely old and incredibly developed mega-civilizations that play a decisive role in the creation and development of our Universe and all intelligent species. If one tries to explain the world and our species not taking into account the part of the mega-civilizations, one will fail to create an adequate picture of reality. Most contemporary scientists refuse to accept the existence and the role of mega-civilizations when explaining the Universe, life, intelligences, and evolution.

Creating a scientific picture of our Universe, life, and intelligence, we should never forget that there are civilizations which are much older than our Universe that not only have the power of the biblical God, but even a much higher creative and control potential, which we can't even comprehend.

Millions of predictions all around the world came true. The fulfilled precognitions are many and often so specific that they are beyond chance probability. But millions of predictions failed. Why? Was it because the psychics were not good enough? Because ordinary people are not good seers into the future?

Most probably it was because they don't have direct access to the simulated future but receive predictions through the System which manipulates the information.

Humans can't know the real future at will.

Why are we receiving true and false predictions?

True information about the future received at will would bring tremendous power (financial, political, social, etc.) in the hands of individuals who possess the gift of precognition. Such people would have absolute control over the world. They could make zillions from future patents, the stock exchange, casinos, foreign currency exchange rates, real estate, oil and gas extraction, all sort of lotteries, investing in the most profitable technologies to be invented, etc. Knowing the future would make some individuals undisputed masters of the world.

People seeing the future and using this knowledge for profit and world control would change history. But this is against the plans of the superior intelligence that is in charge of our world. The world is controlled by the System and the operator who has his own plans.

The System is deliberately spreading true and false information about the future. There is no such thing as 100 percent accuracy of precognitions. We should rely on our sound mind, logic, and science; the fulfilled predictions are only a helping hand we receive sometimes. The System decides when and what real precognitions should be delivered to some individuals. Psychics and mediums are only instruments of predicting the future charade.

Free will can't really exist in a world where the future is simulated and determined.

According to Einstein, there is no such thing as free will. He wrote that he is a determinist.

Arthur Schopenhauer said, "A man can do what he wants, but not want what he wants."

It is possible that in a simulated world, primitive creatures like us have no free will (or a very limited one). But is it probable that advanced future people could have more free will? The evolution and uplifting of underdeveloped creatures like humans is much more effective when their future is simulated and determined.

On the other hand, for our own good, the operator created an illusion of free will.

The following case is from Arthur Walter Osborn's book *The Future Is Now: The Significance of Precognition.*

"Approximately on the first of December, 1923, I dreamt I was in the back seat of a small car. On approaching a crossroad I noticed a big white house on my left, and on my right what appeared to be a hotel, painted yellow with a red sign. No writing was on this sign. When this car was crossing the intersection, a big car, which appeared to be American, smashed into our car, turning it over. The others appeared to be killed; I was not. The dream was fulfilled as follows: On the Sunday after Christmas I went down to the beach for the day, and the dream turned into fact, even to the sign which was in the process of being painted; the only difference was that I told the driver to stop at the comer. At this moment, a big black Buick crossed by at high speed."

Some people believe that some warnings (received in dreams, visions, by mediums and psychics, etc.) are preventing catastrophes, avoiding accidents, etc. Can we change some negative future events? We can't be sure. Most probably the "change of the future" is actually an integral part of the future. Why? The System is playing with us. It is trying to convince us that there is a free will and we can change our future and control our destiny. Only the System can make some changes in our lives. No kahunas, magicians, great political and military leaders, influential authors, philosophers, etc. can change the future.

The question is not whether our life is predestined but to what extent it is predestined. Are all minute details of our life predetermined or are they the result of chance events? We cannot know.

To have or not have free will totally depends on the operator of our Universe. We don't know his plans for the future of humans with respect to free will.

It looks like we have free will, but we don't. It is possible that the System is creating people and a society that in the (distant) future will have free will. Now people are more like remote controlled biorobots creating the technological environment (factories, infrastructure, mining, power plants, etc.), the society, and the science/knowledge that future posthumans will need. We, the people whitout free will, are building the future of man with free will. The process of empowering man with free will be very long and gradual, it will not happen overnight. There will be transitional periods with various degrees of free will, and at the beginning it will be very limited. The man with free will who will be master of his destiny will be born/manufactured somewhere in the distant future. But there always will be more powerful entities that will control all less developed entities. Future humans will be latecomers in the society of the super intelligences that originated unfathomably back in time and that are unfathomably more developed.

But why would a superior intelligence simulate our Universe, controlling our world that way? There are several answers. We are a scientific experiment. We are someone's homework. The entire Universe with all civilizations is just a sophisticated computer game. It is also possible that our patron civilization with the simulation of the future provides safe and fast evolution of the civilizations in our Universe.

Richard Dawkins wrote in his book *The Selfish Gene*, "Survival machines [animals, including man] that can simulate the future are one jump ahead of survival machines who can only learn on the basis of overt trial and error. The trouble with overt trial is that it takes time and energy. The trouble with overt error is that it is often fatal. Simulation is both safer and faster."

Humanity is only one of the numerous civilizations of our Universe. The advanced civilizations become more dangerous for themselves and for the other space races. We already have the technology to destroy humanity, tomorrow we will have the weapons to destroy our planet. In the future numerous novel and much deadlier weapons, technologies, industries, science installations, science experiments, etc. will bring humans and other intelligent creatures to the brink of extinction.

Space creatures, including humans, have to compete not only with their own AI, but also with alien AIs and robots. The AI will create their own technological civilizations which would not only compete the human civilizations but could also be deadly dangerous.

The patron civilization should control a great many civilizations in the Universe that have to compete fiercely; they have to survive, at least most of them. It's a difficult task because the space races have to wage countless wars as a part of the competition, and they have to survive and evolve. Implementing a simulated future is much safer and the development of the civilizations is faster.

The patron civilization is constantly (and clandestinely) introducing new knowledge (scientific, artistic, technological, etc.) into our society.

The development of the sentient creatures, society, sciences, arts, etc. is much better and faster when it is guided by a superior civilization. The intellectual capacity of man is much inferior compared to the capacity of the patron intelligence. Humans can't develop fast enough when they are left alone. If humans have free will and they are writing their history, the development of mankind would be much slower.

Chapter 14
Adolf Fritz: The Surgeon from The Beyond

Andrija Puharich was an American medical and parapsychological researcher, medical inventor, physician, and author. He was licensed for the practice of medicine in California, Maine, and New York, with a specialty in internal medicine. During his scientific career Dr. Puharich held 56 patents for his inventions in the fields of medical electronics, neurophysiology, and biocybernetics.

Dr. Puharich had various academic and professional memberships: the New York Academy of Sciences, the American Association for the Advancement of Science, Aerospace Medical Association, the American Association for Humanistic Psychology, and the American Society for Cybernetics. He published over 50 papers and articles in scientific and professional journals

Puharich investigated Uri Geller, the Dutch psychic Peter Hurkos, the Brazilian healer Arigo, and other persons with paranormal abilities.

In 1963 Andrija Puharich and the businessman Henry Belk visited Brazil to begin an investigation into Arigo's healing powers. Belk had studied psychology at Duke. The Belk Research Foundation had been founded to unite such fields as biology, medicine, physics, chemistry, and other disciplines in the study of the paranormal.

John G. Fuller wrote in his book *Arigo: Surgeon Of The Rusty Knife* about the investigation of Puharich and Belk and the extraordinary life of Arigo.

Guy Lyon Playfair also wrote a book, *The Flying Cow*, 1975, about the psychic wonders in Brazil. The author had spent two years as a member of the Brazilian Institute for Psychobiophysical Research, which investigated and documented a wide range of inexplicable phenomena: poltergeists, psychic surgeons, trance artists, etc. Playfair collected eye-witness evidence for the mysterious abilities of the legendary figure Arigo.

Arigo (pseudonym of Jose Pedro de Freitas) had only a third grade education and no medical training, but he could almost instantly make accurate and confirmable diagnoses or blood pressure readings with scarcely a glance at the patient. He performed hundreds of operations daily with a scalpel, scissors, tweezers, ordinary kitchen knife, or jackknife without antiseptics and without anesthetics, without bleeding control (hemostasis), without tying off blood vessels, without major bleeding. His only concession to cleanliness was to wipe his knife on his shirt before and after surgery. But there is no record of a patient ever having become infected, despite the unsterile conditions. The surgical wounds would heal remarkably fast. Arigo saved the life of patients with cancer and other fatal diseases who had been given up as hopeless by leading doctors and hospitals in Brazil and many advanced countries. He never charged for his services.

Dr. Puharich verified Arigo's healings. He and a cameraman took explicit color motion pictures and black-and-white photos of his work and operations.

Arigo had a thick black mustache, black hair, and a bronzed face. He appeared unshaven and very rustic. But when Arigo entered the small examination room, he seemed to be a totally different person. Now he spoke sharply, like a Prussian officer, with a thick German accent. He was explosively speaking his harsh and guttural Portuguese, sprinkled with words and phrases that must have been German.

The husband of a young Polish woman, being Austrian, spoke to Arigo in German, and was answered in that language.

Arigo claimed he incorporated the spirit of a deceased German doctor, whom he identified as Dr. Adolf Fritz. It was Dr. Fritz, Arigo claimed, who did the operating and the prescribing of the complex pharmaceutical agents he wrote so swiftly. Dr. Fritz was a German physician who had died in 1918.

Arigo would place a patient against the wall, wipe the paring knife on his shirt again, drive it brutally into a tumor or cyst or an eye or ear, and remove whatever the offending tissue was, in a matter of seconds.

There was no anesthesia, no hypnotic suggestion, no antisepsis, and practically no bleeding beyond a trickle.

Puharich and Belk noted that Arigo rarely asked patients a question.

Puharich was watching carefully for hypnosis. It could at least explain part of the procedure. But there was no evidence of it. If anything, Arigo himself seemed to be in a trance state.

Both the Brazilian Medical Association and the Roman Catholic Church were pressing hard against Arigo in a court case and apparently had been scrambling unsuccessfully to find a single case where Arigo had injured someone.

Puharich and Belk had been looking for trickery in watching Arigo, but had to admit that they could not find it.

Andrija Puharich wrote about the operation on his tumor in his book *Uri: A Journal of the Mystery of Uri Geller.*

He rolled up his right sleeve. Arigo asked if any of the surrounding patients could lend him a pocket knife. One man offered a knife, but Arigo said it was too dull. Another man offered a Swiss army knife. Arigo said, "This is a good knife."

Puharich wrote, "All I could feel was something like a fingernail being pressed into the skin. Within five seconds Arigo displayed an elongated egg-shaped tumor for all the patients to see and handed it to me with the pocket knife. I had felt no sensation of pain. When I looked at the wound, there was a trickle of blood from the incision, which was about a half inch long."

Arigo had not washed or disinfected either the skin or the knife.

After the operation Puharich was convinced that Arigo had extraordinary powers in surgery, bacterial control, and anesthesia.

"It is now ten years since the operation. The surgical scar remains; the tumor is still in a bottle of alcohol; there have been no complications."

From 1963 to 1968 Puharich led a number of medical research expeditions to Brazil to study the healer Arigo. Arigo's totally unorthodox surgery and healing powers defied every rule of medicine. Puharich witnessed thousands of Arigo's operations during his investigations.

Arigo died in a car accident in 1971 aged 49.

Later Dr. Juscelino Kubitschek, the former President of Brazil said to Puharich that he had visited Arigo two weeks before his death, and in the most simple and casual way Arigo had said that he will soon die a violent death.

Puharich wrote about the amazing events and the mysterious prophesy concerning the death of Arigo. He got the news of Arigo's death through a mysterious phone call at least fifteen minutes before the event occurred.

The death of Arigo did not mean the end of Dr. Adolf Fritz. Other Brazilian psychic surgeons claimed to be channeling the spirit of the German doctor, including Oscar and Edivaldo Wilde, and the gynecologist Edson Queiroz. The Wilde brothers both died violently in car crashes, while Queiroz was stabbed to death.

Currently, Rubens Farias Jr., a former Sao Paulo engineer and computer programmer, claims to be the channel for the spirit of Dr. Fritz, who has chillingly predicted a violent death for Farias. On a typical weekday, as many as 800 patients will line up outside the hall he uses (on weekends, it serves as a bar), waiting for treatment sessions which might be as short as 30 seconds. While in character as Dr. Fritz, he adopts a German accent and expressions such as "Schnell!"

Did the spirit of the dead German doctor Adolf Fritz really inhabit Arigo and other psychic surgeons? Did he perfectly diagnose patients and do miraculous operations with a dirty knife? How was it possible for a spirit to control the pain without using anesthetics? Could some spirit control bleeding and microbes, preventing infections? As with all stigmata wounds, Arigo's operations never became decayed, infected, or even inflamed.

Arigo was like a remotely controlled robot in the "hands" of the System. He could even speak German when he was addressed in this language. The System had total and precise control over Arigo's mind and body, over the bodies of the patients, and over the microbes.

The spirit of the dead German doctor Adolf Fritz was only one of the many disguises of the System which can control our material world, including us and the microbes.

The System is healing supernaturally sick people through healers like Arigo, shamans, all sort of psychics, and even a number of true doctors with medical diplomas.

But why was the charade with the spirit of the dead doctor Adolf Fritz, the harsh and guttural German accent of Arigo while healing, the Portuguese sprinkled with German words and phrases, the insolent behavior of a Prussian officer talking with great confidence? That way the System convinced Arigo's patients that they were healed not by some uneducated, unshaven, rustic former miner but by a personage who incorporated a supernatural entity which knows everything and can do the impossible.

Chapter 15
Psychic and UFO Healings

In his book *Mind to Matter: The Astonishing Science of How Your Brain Creates Material Reality* Dawson Church reported the next case.

The former cardiac patient Richard Geggie told this story to Church.

Geggie, after having an electrocardiogram and several tests, was told that his heart was at serious risk and his arteries were severely clogged—he should keep nitroglycerine pills with him at all times and not go outside alone.

As a high-risk patient, he was given an immediate appointment for heart bypass surgery.

"I went to the hospital and I was given an angiogram. This involved shooting dye into my arteries through an injection in my thigh. The surgeons wanted to discover the exact location of the blockages prior to the operation.

"When the new angiograms came back from the lab, the doctor in charge looked at them. He became very upset. He said he had wasted his time. There were no blockages visible at all. He could not explain why all the other tests had shown such severe problems."

Geggie later discovered that his friend Lorin Smith, a Pomo Indian medicine man, after hearing of the problems with his heart and arteries, had assembled a group of his students for a healing ceremony. He covered one man with bay leaves and told him that his name was Richard Geggie. For the next hour, Lorin led the group in songs, prayers, and movement. The next day, Geggie was healed.

Who or what had healed Geggie? The shaman Lorin and the students didn't heal him—they were just singing, dancing, and saying prayers. Some powerful entity cleaned the severely clogged arteries, even though Geggie couldn't feel any cleaning. The System has full knowledge and full control over the human body. It can make changes in the human body in a way that individuals cannot feel the changes. It can heal even the most severe diseases.

The following is an excellent example of faith healing.

In 1962 Vittorio Micheli, a soldier in the Italian army, was admitted to the Military Hospital of Verona, Italy with a large cancerous tumor on his left hip.

Because of his severe condition he was placed in a hip-to-foot plaster cast.

His prognosis was so bad that he was sent home without treatment. Within ten months his hip had completely disintegrated, leaving the bone of his upper leg floating in nothing more than a mass of soft tissue. He was unable to move his left leg.

In May 1963, the cast was removed and replaced with a stronger one because Micheli wanted to go on a pilgrimage to Lourdes, a major place of Roman Catholic pilgrimage and of miraculous healings.

On arrival in Lourdes, he lay on a stretcher, required sedatives for pain, experienced anorexia, and felt that his leg was separated from his pelvis. He was immersed in the bath water with the cast on his leg. After the bath his appetite returned and he felt renewed energy.

Thereafter, Micheli was able to walk with the aid of two crutches, and then two "sticks" without any sedatives or analgesics. After the plaster cast was removed in 1964, he realized that his leg functioned normally and he was able to walk without the cast.

Micheli's tumor disappeared, and his bone regenerated. X-rays from 1964 to 1971 confirmed that the bone was reconstituted in structure and the calcification was excellent.

Clinical examinations since the pilgrimage to Lourdes confirmed that he was without pain and walked with a slight limp.

Over the subsequent years, there was no evidence of local recurrence or metastases. Micheli lives a normal family and social life, works at an arduous standing job, and walks in the mountains and plays games.

The next case is from *The Secret Science Behind Miracles* by Max Freedom Long.

J. A. K. Combs, a close friend of the author, attended a beach party. One of the guests fell and there was the characteristic snapping sound of breaking bones.

"Inspection showed a compound fracture of the left leg just above the ankle. The bone ends pressed visibly out against the skin. Combs, who had heard the familiar sound of breaking bones and had himself suffered such a break, realized the seriousness of the injury and proposed that the man be taken at once to Honolulu for treatment, but the elderly kahuna arrived on the scene and took over. Kneeling beside the injured man, she straightened the foot and leg, pressing on the place where the ends of the broken bones pushed out the skin, and then began a low chanted prayer for healing. In a short time she fell silent. Those who stood about watching tensely could see nothing until her hands suddenly moved slightly on the man's leg, and she took them away, saying quietly in Hawaiian, "The healing is finished. Stand up. You can walk.""

The injured man rose to his feet, took a step, and then another. The healing was complete and perfect.

Walter Bromberg, in *The Mind of Man : The story of man's conquest of mental illness*, 1937, wrote that when in trance: "Dull peasants became mentally alert, and could even foretell events or understand things ordinarily obscure to them. Somnambulists made medical diagnoses in other patients brought before them, and foretold the future. The magnetizer of the 1820's merely brought his patient before a competent somnambulist, and waited for the diagnosis."

Some of the best psychics and healers talk a lot of nonsense, and for this reason many people reject their psychic and healing abilities. But actually these psychics are among the best. Since ancient times, people knew that illiterate shepherds were among the best prophets and healers.

This duality is part of a bigger scheme of the System: science and the paranormal should exist and develop side by side. Both are part of the human evolution.

Of course, "dull peasants" can't make medical diagnoses and foretell the future. They receive due information from the System which knows everything about everything and everybody on Earth.

Don't worry about the mumbo jumbo (depending on the century and the beliefs) the healer is saying to you. Now is the time of New Age mumbo jumbo. You will be healed by the System (if the System decides to do so), not by the mumbo-jumboer who is only an avatar of the System.

After absurd statements about visiting alien planets, higher planes of existence, raising your vibration, revealing the "real" culprits for absence of the world peace, finding petrol in a volcanic mountain, and many more absurdities, the psychic healer will diagnose you and prescribe herbal remedy that will indeed cure your allergy that doctors couldn't heal after a long time of swallowing pills and smearing your skin with various medications, some of them imported and very expensive.

Never expect the psychics to be some intellectual titans knowing everything. On the other hand, never expect the doctors, professors, or scholars to be some intellectual titans knowing everything, too.

Never expect them to be honest. Check everything at least thrice. And be ready for more check-ups. Some are incompetent, fake, or tricksters.

Beware of con artists pretending to be psychic healers! Some real healers are also tricksters pretending they can cure diseases that they can't heal. Some medical doctors are doing the same. There are charlatans in all professions. There are dishonest doctors and self-deceiving mystics pretending that they can heal every disease and every illness.

The System urges us to use and further develop both state-of-the-art conventional and alternative medicine. And to do cutting-edge medical research.

The System demonstrates to us that miracle healings exist. On the other hand, the "miracle" healing is a result of the technology of the System. We have to discover the technologies of the System and cure all possible diseases without relying on the System, psychic healers, shamans, angels, saints, etc. to heal us.

The psychic healers (some of them medical doctors), even the best of them, actually don't know how they are healing; they just wave their hands manipulating the body "energy," place needles at specific points in the body, see in their mind what herbs or drugs to use for the given illness, etc. The System shows them what to do but it does not reveal to them exactly what happens during the healing.

Conventional medicine (in most cases) knows what is happening during the healing process, using medicines, operations, and technology.

Al Layne, Edgar Cayce's friend and assistant, learned that the patients who received trance readings did not necessarily have to be in the same room or even in the same city. The next case is an example of diagnosing illness from a long distance. The patient was Bill Andrews of New York. The case is from the book *My Life as a Seer: The Lost Memoirs* by Edgar Cayse.

There are 1469 km (913.6 miles) from New York City to Hopkinsville, Cayce's residence, which takes about 14 hours, 34 minutes to drive.

Cayce diagnosed stomach disorder and prescribed Clarawater.

Andrews wrote back from New York saying that the diagnosis agreed on all points with physicians, but he had been unable to find

Clarawater. He was advertising for it in the *Journal of American Medicine*.

Cayce did another trance reading and provided Andrews with the recipe. He should dissolve in distilled water garden sage, ambergris, grain alcohol, syrup, cinnamon, and Gordon's gin.

Andrews wrote that he had received a letter from Paris from a reader of the medical journal who said that his father had manufactured and sold a product by the same name fifty years ago. He enclosed his father's formula so that Andrews could make it for himself. It was identical to the formula that Layne had obtained in the reading.

The psychics are not "reading" the patient directly, but they receive the due information about him or her from the System. Even if the patient is in the same room, the psychic receives the information through the System which monitors the patient, decodes the information, processes it, and transforms it (in the form of visuals, voice, or thought) so that the psychic can understand it.

Conventional medicine and PSI healing are not contradictory. Conventional medicine still doesn't understand and doesn't use PSI healing. But it is just a matter of time before they merge.

Many researchers write that psychic healing is the result of the placebo effect. But experiments refute such a suggestion.

William F. Bengston and David Kringsley wrote about the psychic healing of cancer in mice in their article "The Effect of the "Laying On of Hands" on Transplanted Breast Cancer in Mice."

After witnessing many cases of cancer remission associated with a healer who used the "laying on of hands," William F. Bengston "apprenticed" in techniques alleged to reproduce the healing effect. The authors obtained five experimental mice with mammary adenocarcinoma, which had a predicted 100% fatality between 14 and 27 days after the injection. Bengston treated these mice for 1 hour per day for 1 month. The tumors developed a "blackened area," then they ulcerated, imploded, and closed, and the mice lived their normal life spans. Control mice sent to another city died within the predicted time frame. Three replications using skeptical volunteers (including David Kringsley) produced an overall cure rate of 88% in 33 experimental mice.

The mice have no minds and opinions, which makes animals useful for studies in which researchers wish to eliminate the placebo effect. Bengston's skeptical students also didn't believe in healing. So it wasn't belief that was producing the healing.

Similar experiments were successfully conducted by other researchers and psychics. They confirmed that psychic healing works.

That eliminates the pseudoscientific claims that the numerous psychic, religious, and UFO healings are a result of a placebo effect. The placebo effect cures are only a small percentage of the healings.

But yes, the placebo effect works, too. Probably!

In their article "A controlled trial of arthroscopic surgery for osteoarthritis of the knee," J. Bruce Moseley et al. reported a placebo surgery.

A group of 180 patients suffering from osteoarthritis of the knee (many were using canes to walk and all those diagnosed were in need of knee surgery) participated in a double-blind experiment in which some were given real knee surgery and others were given placebo surgery. Patients in the placebo group received skin incisions and underwent a simulated removal of damaged tissue without insertion of the arthroscope. The results were startling: all of the patients reported much less pain, and those who had the placebo surgery were even able to walk and play basketball.

But whether the startling healings were a result of the placebo surgery or whether the System cured them, we don't know.

The next case is of supernatural diagnosis. It is from the article "Diagnosis made by hallucinatory voices" by Ikechukwu Obialo Azuonye published in *The British Medical Journal* in 1997.

A healthy woman was at home reading. She heard a distinct voice inside her head that told her that she should have immediate medical treatment because she had a brain tumor and had given her an the address of the computerized tomography department of a London hospital. The scan confirmed the diagnosis. Surgeons removed a meningioma about the size of an egg.

After the operation the voice told her, "We are pleased to have helped you. Goodbye."

Of course, this was not the voice of some spirit of dead doctor but the System which diagnosed the woman and advised her to have a brain scan because she had a tumor in her brain.

In previous times angels, gods, saints, spirits, etc. were miraculously healing people. They used their hands or "divine" light to cure. Now all sorts of UFO aliens and their robots are coming from distant planets or from other dimensions. They diagnose the experiencers with various sophisticated gadgets which often are emitting healing light. The paranormal cures reflect the times in which the patients live.

Now the paranormal experiences become technological—interstellar spaceships traveling from distant stars and galaxies to Earth, alien robots, super sophisticated diagnosing and healing technologies. Modern times and the future are technological. The magical healing of today is the technology of the future.

In his book *The Healing Power of UFOs: 300 True Accounts of People Healed by Extraterrestrials* Preston Dennett reported 300 UFO healings. He wrote,

"After studying the accounts of healings, I began to look at UFOs more as floating hospitals than anything else.

The types of healings seemed straight out of science fiction. People reported their bodies being opened and closed with lasers that left no scars. They told how various organs were removed and put back in again. They reported instantaneous cures of wounds and injuries. They reported healings of serious conditions, such as pneumonia or liver disease. They even reported healings of serious diseases, such as cancer."

Who or what is healing us: the healer, the aliens, or the System? In some cases there is no healer at all—bathing with holy water, witnessing UFOs, hallucinatory alien doctors or angels. In such cases the System is the healer. Healing from a distance by a healer is not possible. The System, not the healer, is curing sick patients. But in the presence of a healer, a shaman, a saint, or an angel, people know who is healing them. He is a wise man; he makes prophecies, gives advice. Through them the System is socially controlling the people, their way of thinking, their actions.

The supernatural healings are not a result of some magic but of the System's technology.

How does the healer know specific medical terms? Who or what is telling him the medical diagnosis? Obviously the System knows the medical conditions of all people on Earth, knows perfectly contemporary (and future) medicine and drugs, and is controlling the mind of the healer. It can heal all kinds of diseases. Why isn't the System healing all people on Earth that way? People have to develop science, technologies, and medicine in order to cure themselves every time they get sick, and not to expect the System (disguised as angels, gods, ufonauts, saints, psychic healers) to heal them, when it decides. Future people should take control over their lives. They have to become masters of their bodies, lives, and destinies. The future of man is technological, not spiritual and not "magical." What to us now seems like magic (and magical healing) is actually advanced technology created by the System and the advanced civilization behind the System.

Edgar Mitchell, Apollo 14 astronaut, said, "There are no unnatural or supernatural phenomena, only very large gaps in our knowledge of what is natural. We should strive to fill those gaps of ignorance."

Chapter 16
Dinosaurs on the Moon or Engineered Extinction

Historic and evolutionary events we think are accidental and natural could actually not be that accidental and natural, but caused by a superior civilization. We still can't tell for sure what is natural, what is artificial, what is accidental, and what is manipulated. Alongside the natural evolutionary activities, many of the processes (geological, evolutionary, genetic, cultural, historical, etc.) could be assisted.

Dr. Dale Russell of the National Museum of Natural Sciences in Ottawa, Canada, coined the word dinosauroid, an intelligent creature that evolved from the dinosaurs. He claimed that some dinosaurs had all the ingredients for success that we see later in the development of apes, and that they were well on their way to becoming a sentient species.

Several dinosaur species were very manlike: they stood around two meters high on their two hind legs and had a relatively large braincase, stereoscopic vision, and hands with opposable thumbs. Their forelegs with three slender, flexible fingers were ready for use as hands. Some dinosaurs were nearly warm-blooded, an important step toward intelligence.

The hypothetical dinosaur civilization could have started on Earth during the Cretaceous period or earlier, and would have had at least 66 million year head start on humans.

In recent years, the idea that intelligence history on Earth would have been very different without the extinction of the dinosaurs has become a very popular belief. But did the dinosaurs have all of the attributes considered necessary for intelligence in the intelligent mammal? Was it really possible for dinosaurs to outsmart mammals, dominate our planet, and start colonizing the galaxy 66 million years ahead of us? According to many recent articles and books written by academic scientists, independent researchers, and nonprofessionals—yes!

But there is a very big problem with a supposed dinosaur civilization at the end of the Cretaceous: the method of breeding.

Because of the extensive fossil record of extinct dinosaur eggs, eggshells, and embryos, it is well established that dinosaurs laid eggs, and like most living reptiles and birds, built nests.

The eggs are hatched outside the maternal body.

The principal disadvantages of dinosaur reproduction compared to mammalian are:

1. The nutrients inside the egg are very limited compared to the continuous supply that mammals receive inside the womb;

2. The oxygen supply is much lower as well;

3. The temperature of the reptile embryo is dependent upon the environment, while the body heat of the mammalian fetus is constant;

4. Dinosaur newborns don't get the mammalian highly nutritious food—milk.

The developing sophisticated brain needs more oxygen, more nutrients, a constant temperature, and more time.

The mammalian fetus develops inside the maternal body and can receive the continuous, generous supply of oxygen and nutrients needed to build a complex brain. The milk of mammals contains all essential nutrients, important antibodies, and white blood cells. This is a perfect food for infants and for their energy-hungry developing brains.

Mammals are born in a much more advanced state than egg-laying animals.

Eggs hatch between 60 and 105 days after they are laid. The human baby develops inside the mother's womb over about 266 to 270 days. The human brain develops from three to four and a half times longer in a much better inner environment than the dinosaur brain. The mammalian fetus and newborn get food that is high in nutrients for their growing, unfolding brains.

Even the warm-blooded dinosaur descendants, the birds, are famously unintelligent compared to mammals. Birdbrain is a byword for a stupid thing.

Warm-bloodedness does not help much toward intelligence if one hatches from an egg.

In short, the brain of live birth mammalian animals is evolutionarily higher than the brain of animals that reproduce through egg-hatching, and it is also far more sophisticated. The dinosaurs laid eggs, and their brains couldn't develop enough to outsmart mammals. Thus, the dinosaurs couldn't land on the Moon in the Cretaceous period.

The Cretaceous mammals were an evolutionarily higher species than the dinosaurs and their successors. They had much greater potential, and life on Earth proved that. But their advent was a result of an unlikely event.

Complex life can survive only on planets that experience rare or no celestial impacts by large asteroids and comets that could trigger severe mass extinctions or wipe out almost all living creatures.

On the other hand, the origins and evolution of humankind required an extremely unlikely event such as the Cretaceous–Paleogene (K-Pg) extinction 66 million years ago that removed the dinosaurs as dominant terrestrial animals.

Dr. Robert T. Bakker, a paleontologist at the University of Colorado, said to *The New York Times* in 1990, "It is as if nature aimed a smart bomb at the animal kingdom, designed to kill off only certain groups, particularly the large land animals."

"Why the small animals survived while the larger ones became extinct remains a riddle."

There are about 150 dinosaur extinction hypotheses. What was that smart bomb that killed off the dominant dinosaurs but spared the small, fragile mammals?

In the literature, the K-Pg catastrophe is presented as an accidental slam of an asteroid into Earth that caused the almost instantaneous extinction of most species on the planet, including the dinosaurs, but not the tiny shrew-sized mammals.

The dinosaurs hunted other dinosaurs and animals or scavenged dead animals. Most, however, ate plants like evergreen conifers (pine trees), ferns, mosses, horsetail rushes, cycads, ginkos, and in the latter part of the dinosaur age, flowering (fruiting) plants. Toward the end of the Cretaceous period the fertile soils and the grasses became widespread and there was enough food for the future growing population of mammals. The time was ripe to delete the dominant but outdated dinosaurs and replace them with the new masters of the planet, the mammals.

A large and very vocal group of folks from the academic and entertainment businesses are convincing us that the dinosaurs were destroyed by an accidental asteroid and that such a space body can annihilate humans.

The asteroid theory by Walter Alvarez can't explain many specifics of the Cretaceous mass extinction, like the presence in the soil of extraterrestrial amino acids for tens of thousands of years *before* and *after* the impact; the loss of part of Earth's atmosphere; the multiple layers of iridium; the extraterrestrial soot in the boundary layer; the homogenous distribution of iridium all over the surface of the planet, and other peculiarities.

Researchers reported that they had found isovaline and aminoisobutyric acid deposited for tens of thousands of years *before* and *after* the Cretaceous catastrophe. In the boundary clay itself there are no such amino acids. Some meteorites are rich in organics, but how were these amino acids delivered constantly from space for such a long period of time (tens of thousands of years), and what is the connection with the dinosaur extinction? Why are there no such extraterrestrial amino acids in the boundary clay itself?

The hypothetical Chicxulub asteroid couldn't deliver extraterrestrial amino acids for tens of thousands of years before and after the impact.

The iridium enrichment, supposedly a key proof for asteroid impact, has problems, too.

Researchers are reporting findings of locations where iridium is deposited in more layers than only one, as it should be in the case of an asteroid impact. Were there multiple asteroid impacts? Then where are the craters of such impacts? And what delivered the iridium and the extraterrestrial amino acids between the impacts?

For example, Lattengebirge, in the Bavarian Alps, has three iridium anomalies, below and above the K-Pg boundary. The oldest anomaly antedates the boundary by 9,000 to 14,000 years.

According to Dewey McLean, the following sites have multiple iridium layers: Nanxiong Basin, South China—6 spikes, Braggs, Alabama—3 spikes, Brazos River, Texas—2 spikes, El Kef, Tunisia—2 spikes, Beloc, Haiti—2 spikes.

The distribution of iridium anomalies provides evidence of episodic iridium-delivering events over an extended period of time.

One more problem: the iridium anomaly at the K-Pg boundary appears to have been spread homogeneously all around the globe. Iridium should decrease with increasing distance from the impact site until it is altogether absent, if it was an asteroid collision.

Jason Moore and Mukul Sharma from Dartmouth College in New Hampshire compiled all published data on iridium and osmium amounts from the boundary layer. In their final analysis, the overall iridium and osmium levels were much lower than those that scientists had been using for decades, which indicates a smaller impactor. "But an asteroid that size would not make a 200 km-diameter crater," wrote Moore.

So what caused the mass extinction and the huge crater, if it wasn't an asteroid?

On the other hand, there were also other asteroid impacts that created huge craters; however, there were no extinctions at all.

The hypothetical Chicxulub-sized asteroid was too small to cause the mass extinction.

An asteroid the size of the Chicxulub space bolide can't cause a mass extinction despite the claims of so many scientists, futurists, science fiction writers, and moviemakers. Yes, it would cause tremendous damage, but the dinosaurs and many other species would survive and thrive.

So, what could cause such a huge crater and the mass extinction, if it wasn't an asteroid? What could deliver amino acids and iridium from space before and after the impact?

Numerous scholarly articles have reported high contents of soot in the boundary-clay layer. Asteroid theory says that wildfires were ignited by the impact fireball, whereas globally they were ignited by radiation from the reentry of hypervelocity ejecta.

Here, the problem is that most of the soot in the boundary layer is specific, as if created by burning petrol or other carbohydrates. What was that mysterious material that burned and contributed to the large part of the soot that covered the entire surface of the Earth? Some interpretations have suggested that the soot from carbohydrates came from the combustion of fossil fuels such as crude oil, coal, or oil shales near the Chicxulub impact site. But the homogenous distribution pattern over the planet's surface and the composition of the soot do not support such a hypothesis.

The asteroid hypothesis is very nice, simple, and sexy, and very cinematic, but it can't explain in a satisfactory way all the specifics of the massive Cretaceous die-off.

The picture of the Cretaceous catastrophe is too rich and complicated for such a simplistic picture provided by the asteroid advocates.

Only the K Comet theory (K for Cretaceous) can explain all the peculiarities of the Cretaceous-Paleogene extinction.

Dinosaurs could survive the Chicxulub asteroid, but they couldn't survive the K Comet events.

The likely source for the elemental carbon and extraterrestrial soot from burning hydrocarbons in the boundary layer is a comet.

Comets contain large amounts of frozen liquids, gases, and organics, which burn like giant fuel bombs while entering Earth's atmosphere, leaving behind huge quantities of extraterrestrial soot. Frozen gases and liquids like ammonia, methane, ethane, acetylene, methyl alcohol, and many other chemicals have been seen in varying abundance in comets.

There are significant amounts of frozen hydrocarbons in comets. Comet Hyakutake, the brightest comet in 20 years, had a big surprise: 50 million tons of frozen ethane, a hydrocarbon common in crude oil. Chemical analysis showed that the abundances of ethane and methane in the comet were roughly equal. Comet Hyakutake could deliver no less than 100 million tons of burning and detonating hydrocarbons if it entered Earth's atmosphere.

Most of the soot and carbon at the end of the Cretaceous period was delivered by a burning comet and cometary dust. In the boundary clay there is also some amount of soot from wildfires.

In 1980, *Science* magazine published a dinosaurs-killed-by-a-giant-asteroid theory by Luis Alvarez. Critics asked how creatures outside the impact area were killed. Alvarez replied:

"From darkness. The impact created huge amounts of dust, cutting off the Sun's power by up to 20% for 8 to 13 years."

Actually, the "dark times" lasted much longer, tens of thousands years, and started a long time before the impact events.

The isovaline and aminoisobutyric acid tens of centimeters below and above the K-Pg boundary were dusted by a comet.

According to the research data, the entire *boundary zone*—a roughly 100-cm layer encompassing both sides of the boundary—which represents the catastrophic events, was formed over about 20,000 to 100,000 years.

The thin 1-cm boundary-layer clay is approximately in the middle of the boundary zone.

Long-period comets have highly eccentric orbits, extending to the far reaches of the Solar System.

Sometimes they make passages close to the planets and the Sun, diverting into the inner Solar System and becoming short-period comets. Short-period comets have orbital periods of less than 200 years.

When comets approach the inner Solar System and the Sun, they begin to sublimate (cometary material transits directly from solid state to gas) and vaporize, creating an envelope of thin gas and fine dust called coma.

Sunlight pushes away the gas and dust of the comet to form a tail.

A comet is exhausted or extinct when most of the volatile material contained in the nucleus is evaporated away by the Sun, and the comet becomes a much smaller, dark, inert lump of rock or rubble that can resemble an asteroid.

Short-period comets in the inner Solar System typically have a lifespan of only thousands to tens of thousands of years.

If large short-period comets don't get close to the Sun, their lifespan is much longer—about 100,000 years.

Probably the size of the original K Comet was about 100 km in diameter. The K Comet became a short-period comet with a life span of about 100,000 years or less.

There are now 17 currently active Earth-crossing comets. The term Earth-crossing comet means a comet is on an orbit that, as a consequence of perturbations, can intersect the orbit of the Earth. And, of course, hit our planet. The K Comet was nearly on a collision course with Earth.

There are several peaks of iridium and extraterrestrial amino acids before and after the boundary layer. They possibly could be caused by fragmentation of the K Comet, hence the increased levels of cometary dust or/and airbursts (explosions in the air). Smaller fragments of the comet could hit the land or the ocean before, during, and after the main impact.

The K Comet's leftover fragments remained moving in Earth-crossing orbit until they were exhausted—tens of thousands of years is a usual time for such a comet to become fully exhausted.

If the orbit of the cometary dust intersects Earth's orbit, our planet and its atmosphere will sweep through the dust stream every year, experiencing meteor showers and the deposition of fine dust on the surface of the globe.

Because some of the dust particles are very small, they will be rapidly slowed to a stop in Earth's upper atmosphere. Instead of burning up in a flash of light like the larger cometary grains, they will drift slowly to the surface of the planet. It will take months or longer for fine cometary dust to settle down from the upper atmosphere.

In such a flyby of a huge comet, Earth would accumulate a large mass of dust in the upper atmosphere, changing the climate and somewhat inhibiting the photosynthesis of land and marine plants. There wouldn't be an inky darkness at noon, just prolonged periods that would look like very dark days.

But what would the effect have been on plants and animals from the lower light intensity and a slightly cooler climate?

Major food chains were disturbed. The reduction of the plant mass led to the starvation of plant-eating animals. The first victims were the large herbivores on land and in the oceans, especially the ones living at the Polar Regions, where the sunlight reduction by the dust cloud was more serious, the temperature drop was substantial, and the loss of plant mass was significant.

Large species at the top of the food chain, such as dinosaurs, are highly vulnerable to ecosystem disruption.

The dust clouds phase of the K Comet events lasted for tens of thousands of years before and after the Chicxulub cometary impact; thus, the Cretaceous extinction began tens of thousands of years before the catastrophic impact events.

The end-Cretaceous cooling of the climate is confirmed by many researchers.

Jason Moore and Mukul Sharma claim that the iridium and osmium levels across the K-Pg boundary indicate a small impactor, and an asteroid that size would not make a crater 150-200 kilometers in diameter.

Asteroids travel too slowly for a small rock to generate enough energy to create the Chicxulub crater. Comets travel a lot faster than asteroids, and a cometary fragment of about 7 kilometers across, traveling at typical comet velocities, could release enough impact energy to create the Yucatan crater, suggested Moore and Sharma.

The kinetic energy of an incoming object from space follows the equation: $Ek = ½ mv^2$.

Ek = kinetic energy, m = mass of object, v = velocity or speed of object.

An object moving at twice the velocity of another object with the same mass has four times the kinetic energy.

The K Comet was large and very fast, and the consequences for terrestrial life were tremendous. But even before the catastrophic impacts themselves, it started to kill off the terrestrial biota because of the cometary dust cloud.

Large space bodies strike the ground with a significant fraction of their cosmic velocity. The kind of crater and degree of destruction depend on the size, velocity, composition, degree of fragmentation, and incoming angle of the impactor. New computer simulations by an international team of researchers from Imperial College London suggest the asteroid (actually the fragment of the K Comet) that doomed the dinosaurs 66 million years ago struck Earth at the deadliest possible angle.

Impacts of such a disintegrating huge comet could cause colossal earthquakes, giant tsunamis, massive wildfires of plants and fossil fuels all around the globe, and tremendous hurricanes, and might activate volcanoes and basalt floods, changing the chemistry of the oceans. The skies would be covered with a thick dust blanket.

A burning, disintegrating huge comet can create a heat wave in the atmosphere with a devastating effect on the biota that cannot be created by an a stony or iron asteroid.

The heat impulse lasted at least several days and destroyed a large proportion of the plants of many regions without burning them. The vegetation was destroyed by the thermal impulse, acid rains, numerous local wildfires, and the lack of enough sunlight.

Most of the cometary aminoisobutyric acid and isovaline could not survive the fiery impact (especially taking into account that cometary material partially consisted of combusting stuff), so researchers didn't find these amino acids in the boundary clay.

Over millennia, the whole chemistry of the Earth changed, also due to the organisms, which appeared about 3.8 billion years ago. Oxygen, which is released as a by-product of photosynthesis, appeared in Earth's atmosphere; carbon dioxide was depleted over the ages.

Atmospheric pressure during the Mesozoic period was about 3 to 8 bars, and it was declining steadily. The oxygen level was getting higher—it was between 24 to 28 percent; some researchers state higher or lower amounts.

Now the air in Earth's atmosphere is made up of approximately 78 percent nitrogen and 21 percent oxygen. Air also has small amounts of lots of other gases, too, such as carbon dioxide, neon, and hydrogen.

Among modern flying birds, the wandering albatross has the greatest wingspan, up to 3.5 meters; the trumpeter swan has the greatest weight, up to 17 kilograms.

Some flying reptiles from the Jurassic and Cretaceous periods weighed about 70 to 130 kg and had wingspans from 10 to 17 meters. The laws of aeronautics and physiology would not allow the huge Mesozoic creatures to fly in the present air, but they ruled the ancient skies because the atmosphere was denser and richer with oxygen.

In the specific Mesozoic hothouse world, with a much denser atmosphere and high amounts of oxygen and carbon dioxide, animals and plants grew much larger and were more numerous.

The huge reptiles and insects could fly only in a dense atmosphere with higher amounts of oxygen. They needed more fuel (oxygen) for their metabolic engines and thicker air to support their wings.

The amounts of oxygen available to the metabolism of Mesozoic animals depended not only on the percentage of this gas in the atmosphere, but also on the air pressure. Higher pressure also means more available oxygen.

The higher temperatures and the higher pressure made the utilization of oxygen much easier.

The breathing system of the dinosaurs and their hemoglobin were adapted to much higher levels of oxygen and a denser atmosphere. They were different than in modern animals.

About 80 to 90 percent of the metabolic energy of animals comes from oxygen and only 10 to 20 percent from food.

The Mesozoic species, especially the dinosaurs, took advantage of the large amounts of oxygen, the abundant food, and the steady, warm climate, with only slight seasonal variations.

The large dinosaurs did not need to be truly warm-blooded because they had enough energy (lots of oxygen and food) at their disposal, a steady, warm climate, and almost no rival species outside the dinosauria.

Not being truly warm-blooded was a way for them to avoid the overheating of their huge bodies in the hot Mesozoic climate. The removal of body heat is more difficult in a hotter and denser atmosphere. The large dinosaurs would have been in trouble, if they were truly warm-blooded.

Mammals, the present dominant species, can't reach the giant size of the Mesozoic dominant species, the dinosaurs, because the modern atmosphere is different—a lower percentage of oxygen and lower air pressure.

Dinosaurs were very well adapted to the Mesozoic period. They ruled over a specific world.

Dinosaurs couldn't live in the present world for many reasons—different atmosphere, different microbes, etc. Thus, present-day dinosaurs would need to be genetically modified in order to survive in the contemporary ecosystem. It's not possible to reconstruct in an open habitat the original authentic animals of the Mesozoic world, as Michael Crichton did in his novel *Jurassic Park*.

What made dinosaurs dominant became their major drawback during the K Comet events.

For the Mesozoic animals, the Cretaceous catastrophe was a metabolic disaster.

During the K Comet events, oxygen in the air decreased abruptly. Because of the tremendous amounts of inflammable volatile elements in the comet (frozen gases and liquids), the very high velocity, hitting in the shallow waters of the ocean, the impact plumes and the column of superheated gases ejected part of the atmosphere into space. With a partially lost atmosphere, the air pressure became lower.

Part of the atmosphere was lost in space; part of the oxygen burned during the impact events.

The oxygen in the deflated post-impacts atmosphere was decreasing because most of the oxygen-producing plants were annihilated by wildfires, the tremendous global thermal pulse, heavy acid rains, the impacts, and the acidification of the oceans. The lower sunlight levels due to massive dust clouds from the impact blast, volcanoes, prolonged fossil-fuel fires, cometary dust, and wildfires reduced the oxygen produced by the dwindling land and marine plants. There was a great loss of phytoplankton, kelp, and algal plankton, which produce about 80 percent of atmospheric oxygen.

After the impact, vegetation suffered a short but severe crisis. The devastation of forests and other plants after the K-Pg boundary impact was a global phenomenon. Half of the plant species died off during the K Comet events and were replaced by other species, which caused additional problems for many animals that were not used to such vegetation.

The energy amounts available to Mesozoic animals during the catastrophic events were tremendously reduced because of the huge loss of plant mass and the drop of available oxygen. The ecosystem could no longer sustain such a great number of animals. Especially affected were the huge species, which could not survive the energy deprivation.

The environmental conditions were the worst during the cometary hits, but the reduced sunlight, the cooler climate, the lower oxygen levels, and the food supply reduction lasted for tens of thousands of years after the impacts.

In the heavily stressed environment during the K-Pg catastrophe, the animals needed even more energy from oxygen and food to survive.

Dinosaurs abruptly lost their metabolic advantages during the catastrophic events because air pressure and oxygen levels dropped, food became scarce, the temperatures dropped, and seasons appeared.

Researchers often ask the question of why some species died off while others survived. The most important factor was body size—only small species survived the harsh period of severely reduced energy. Small animals cope much better with low amounts of food and oxygen. Experiments also prove that small animals perform better in a low-oxygen environment.

The average body size of animals after the K-Pg events was between 2 and 5 kg. With the exception of some ectothermic species such as the leatherback sea turtle and crocodiles, *all tetrapods (four-limbed vertebrates) over 25 kg died off.*

The average surviving animals were as large as cats, chicken, and rabbits. The largest ones were as "huge" as dogs and goats.

The smallest dinosaurs were mainly from the late Triassic and early Jurassic. Most of them perished before the end of the Cretaceous period. Dinosaurs got largest in the late Jurassic and Cretaceous periods.

Large animals couldn't squeeze through the K-Pg energy filter. They were too large. The typical dinosaur body mass was between one and ten tons, far higher than the maximum "permitted weight" of 25 kg. There were of course some small end-Cretaceous dinosaur species, but they couldn't survive the fierce competition for food during the severe period after the comet catastrophe.

After the K Comet catastrophe, the Mesozoic world was over.

Mammals were evolutionarily better players and won the world dominance trophy by a single stroke, thanks to the K Comet. The dinosaurs were simply too large and too Mesozoic to survive the Great end-Cretaceous Energy Filter.

Peter Ward and Donald Brownlee wrote in their book *Rare Earth*, "One could argue that even in the last 500 million years, the time of complex animals, there should have been enough comet or asteroid strikes to exterminate animal life on this planet. That has obviously not happened."

Why had no space body impact exterminated or even threatened complex life in the last 500 million years? Sheer luck for humans and other species, or we are a protected species? On the other hand, for 500 million years there was only one huge celestial impact and it was not just to kill off animals but to replace at the right time the outdated dominant dinosaurs with the evolutionary higher mammals.

The simulations from Imperial College London have revealed the bolide that doomed the dinosaurs struck Earth at the deadliest possible angle. More sheer luck, or was it by design?

At present, science still can't tell if the K-Pg catastrophe was an accidental event or an astroengineering project to steer evolution on Earth.

If one delves deeply into the research papers about the dinosaur extinction, their demise looks more like a very precise extermination operation of these species than an accidental slam of a celestial body into Earth.

This dinosaur extinction is an example of a possible evolutionary control by the System which delivered a large comet in a specific orbit around the Earth. Without the System we couldn't become humans and develop civilization.

Chapter 17
Dowsing

Dowsing is a type of divination employed in attempts to locate ground water, buried metals or ores, gemstones, oil, missing jewelry, architectural structures and artefacts, missing persons (incl. dead people), and many other objects and materials without the use of scientific apparatus.

Dowsing and other forms of divination have been around for thousands of years.

Bas-reliefs from early Egypt show figures carrying at arm's length in front of them a forked stick. Chinese Emperor Kwang Su is depicted in a statue dated 2200 BC carrying an identical object.

Traditionally, the most common dowsing rod is a forked (Y-shaped) branch from a tree or bush. The two ends on the forked side are held one in each hand with the third (the stem of the Y) pointing straight ahead. The dowser walks slowly over the places where he suspects the target (for example, ore or water) may be, and the dowsing rod dips, inclines, or twitches when a discovery is made. Many dowsers today use a pair of simple L-shaped metal rods. One rod is held in each hand, with the short arm of the L held upright, and the long arm pointing forward. When something is found, the rods cross over one another making an X over the found object.

Dowsing is an ancient technique for finding things, solving problems, health diagnose, finding out the sex of an unborn baby, etc.

Occult practitioners, mystics, and common people employed in dowsing have often attributed these effects to a supernatural force.

For thousands of years finding ores and producing metals was close to magic.

In his book *History of Religious Ideas* Mircea Eliade wrote, "The metallurgist, like the blacksmith and, before him, the potter, is a 'master of fire.'... in archaic societies, smelters and smiths are held to be masters of fire, along with shamans, medicine men, and magicians."

But how is dowsing scientifically explained?

One explanation states that electric fields effect human nervous systems and some people can sense electric fields. For instance, moving water, flowing through the Earth's magnetic field, creates its own electric field.

Another explanation suggests that the Earth's electro-magnetic field is deformed by underground water or metal objects and some people can sense these deformations of the electro-magnetic field. The function of the rods is to amplify the very small muscle twitches. The minuscule muscle movements are caused by subconscious mental activity. It looks and feels as if the movements are involuntary.

Scientists have found electric and magnetic sensing organs in many creatures: bacteria, sharks, fish, birds, etc. They, however, didn't find similar structures in humans.

Some dowsing practitioners suggest that they respond to Earth "rays."

Another explanation claims that dowsing is the ability to tune in to some force or unknown energy that science has not yet identified, or dowsing is simply a way of tuning in to the "quantum field."

But the mysterious ability of the dowsers to attune their instruments (dowsing rods or pendulums) to electro-magnetic fields or some still undiscovered form of energy that is emanated from ground water, buried metals or ores, or other hidden objects can't explain health diagnose, finding out the gender of the unborn baby, or map dowsing. Some dowsers or psychics are able to locate things by dowsing on a map with a pendulum. The dowser can discover underground water, oil, metals, etc. by moving a pendulum over a map or a hand-drawn sketch without being physically present on site. In some cases the dowser or the psychic knows where the water is just looking at the map or the plan of the area, not using any equipment like a pendulum or dowsing rods. In the case of map dowsing, the dowsers are not responding to the physical terrain or electro-magnetic fields.

Some psychics do not need auxiliary aids, such as a pendulum or dowsing rods. They receive information in their minds as a voice, feelings, thoughts, visions, etc.

In 1928, Abbé Bouly said in a lecture, "I no longer require a rod, I can see the stream with my eyes; I attune my mind; I am looking for lead, I fix my eyes; I feel a wavy sensation like hot air over a radiator; I see it."

South African Pieter van Jaarsveld became quite rich back in the 1950s, asking £25 to point out the place to drill.

He used no sort of dowsing rod but was able to see water "shimmering like green moonlight" through the surface of the soil. He could follow underground water like that, even in the dark.

Most probably, the subconscious mind can receive information about the physical world and the hidden water or ores through the System which has a huge database of the Earth and its resources.

Although considerable research has been done on telepathy, dowsing, and other similar phenomena, their status still remains uncertain for science. There is a list of prizes offered to anyone who can provide scientific evidence of paranormal abilities. Many people claimed the prizes, but still no one got one. This is because behind the channel is an intelligent entity that makes decisions about what and when to deliver information. The System follows its own plan and strategy. Dowsing and other phenomena we consider paranormal are not controlled by man but by the System. Its strategy is to use some ancient activities we call paranormal to introduce various knowledge clandestinely into human society, and at the same time, to make science a powerful tool for people. The paranormal and science should both exist and coexist together.

Chapter 18
Transformation: Entering a New World or a Hell

A few words about transformation.

The greatest minds of all centuries are transformed individuals. Transformation is the final part of the individual's growing up. Not all people get transformed. Only the chosen ones, chosen by the System.

In his book *Mothman Prophecies* John Keel wrote about a transformation after UFO or religious encounters.

"Persons caught in these [UFO] beams undergo remarkable changes of personality. Their IQ sky-rockets, they change their jobs, divorce their wives, and in any number of well-documented instances they suddenly rise above their previous mediocre lives and become outstanding statesmen, scientists, poets and writers, even soldiers. In religious lore, being belted by one of these light beams causes "mystical illumination." When Saul, a Jewish tent-maker, was zapped by one of these beams on the road to Damascus it blinded him for three days and he was converted to Christianity on the spot and became St. Paul."

"Isolated individuals on lonely back roads will still be getting caught in sudden beams of energy from the sky, then shuck their families, quit their jobs, and rocket into notoriety or plunge into the hell of insanity and bankruptcy."

The UFO beams and the beams from the Bible are only the demonstrative part of the transformation. They and many other mystical manifestations are intended to show that something very important happened to some individuals and they are not the same persons anymore. They are already different, "spiritual," something much more than they were previously. It isn't the light beams that transformed the individuals but the System. The beams and the accompanying mystical, paranormal, or UFO experiences are some sort of demonstration and initiation: admitting someone into a secret or obscure society or group, typically with a ritual.

You cannot transform yourself as thousands of self-improvement books are trying to convince you. The System is transforming you. No matter what transformative practices or techniques you do (yoga, tai chi, various religious practices, occult rituals, mediation, etc.), if the System does not transform you, you will never be transformed. The transformative practices will only help the process, they do not transform.

The transformation has many levels, implications, and manifestations.

The individual will be transformed sometimes quickly, sometimes very gradually, sometimes the process lasts for years, but in most cases the transformation significantly and permanently changes the person.

What actually happens during transformation we do not know.

Not only the mind is transformed, but mainly the energy system (we don't know what it is and whether it is energy at all) that permits the human body, and the human body itself.

One should be very careful. The process of transformation could be dangerous, with negative or even catastrophic results. Do not push, the System will guide you, it makes the decisions.

The System is transforming you not for your amusement; you are employed by the System for a certain job.

Some individuals know that they are in the process of transformation. But in some cases the individuals don't know that. They go through a period (lasting months or years) of mental, physical, and psychological suffering and after that they become much better professionals. They see some changes in their thinking: they are more intuitive, they understand complicated scientific texts more easily, their mind is somehow working better. Usually they don't go through the phase of siddhis or paranormal experiences.

Sometimes the transformative experiences may be distinctly mundane. Transformations can be fast or gradual, mystical or mundane.

What is it like to be transformed? In most cases it means gaining some paranormal abilities; one can heal, read people's minds, see future events, have out-of-body experiences, experience mystic oneness with the Universe, contact at will telepathically with some of the disguises of the System (this does not mean that you will have access to the entire knowledge of the Universe), etc. Your worldview will be changed, too. You will not become some intellectual titan knowing everything but you will be a better professional. You will rise above your previous mediocre life and become outstanding. You will work in cooperation with the System and it will help you in many cases. Many transformed persons talk and teach New Age mumbo jumbo. It is part of the game.

The more picturesque, emotional, and mystic transformational experiences mean a lower level of transformation, low level adepts. The kindergarten rooms are more picturesque and colorful than the study room of a true researcher.

During the transformation process one often gets siddhis. Some siddhis are transient, other paranormal abilities stay. They can't be controlled.

Siddhi is a Sanskrit word that literally means "accomplishment," "attainment," or "success." It is also used as a term for spiritual power (or psychic ability). According to the Eastern traditions, these siddhis are clairvoyance, levitation, knowing the past, present, and future, absolute lordship, the power to subjugate all, become as small as an atom, materialize objects, have access to memories from past lives, one's soul can enter into the body of some other person, resurrection of dead people, increasing or decreasing the size of the body, acquire immense wealth, transform one substance into another, become infinitely heavy or almost weightless, unrestricted access to all places, realize whatever is desired, hearing things far away, seeing things far away, moving the body wherever thought goes, assuming any form desired, dying when one desires, remaining unconquered by others, etc.

Many schools of thought state that the siddhis are a normal set of occurrences that should not be focused upon because they will pull one from the path. According to the Eastern traditions, the ultimate goal is Moksha. Moksha means Nirvana or the liberation from worldly suffering and the cycle of rebirth and reincarnation, leading to an enlightened relationship with the creator.

Mircea Eliade wrote about the transformation in his book *Shamanism: Archaic Techniques of Ecstasy*. The shamans are transformed people.

"More or less pathological sicknesses, dreams, and ecstasies are, as we have seen, so many means of reaching the condition of shaman."

"But usually sicknesses, dreams, and ecstasies in themselves constitute an initiation; that is, they transform the profane…into a technician of the sacred."

The change is profound. Even the vocabulary of the shaman is different from that of his tribesmen. The vocabulary of a Yakut shaman contains 12,000 words, whereas the ordinary language—the only language known to the rest of the community—has only 4,000 words.

The old shamans transform the profane chosen one into a technician of the sacred? Well, the truth is that the transformed individuals (no matter of their further profession and career) become technicians of the System.

"According to another Yakut account, the evil spirits carry the future shaman's soul to the underworld and there shut it up in a house for three years (only one year for those who will become lesser shamans). Here the shaman undergoes his initiation."

"There a giant fir grows, with nests in its branches: The great shamans are in the highest branches, the middling ones in the middle branches, the least are low in the trees. Some informants say that the Bird-of-Prey-Mother, which has the head of an eagle and iron feathers, lights on the tree, lays eggs, and sits on them; great, middling, and lesser shamans are hatched in respectively three years, two years, and one year."

The transformation lasts one year for the lesser shamans, two years for the middling shamans, three years for the great shamans. As you can see, the transformation is not instantaneous as many authors write but a long and often not very pleasant process.

After the transformation some individuals are "adjusted" many times by the System during their entire life. These individuals experience during the "adjustment" physical and mental discomfort, sometimes illness.

Not all mystical or highly emotional experiences are leading to transformation or are a sign of transformation. Changing your worldview after a mystical experience is not a sign that you are transforming. Mystical experiences occur much more often than experiences that are result of transformation. High level transformations are exceptionally rare.

The spiritual teachers promise you eternal joy, love, and happiness. But Ecclesiastes says, "in much wisdom is much grief, and he who increases knowledge increases sorrow." One has to choose—to be "spiritual" or a researcher on the path of wisdom.

There is a great variety of transformed individuals: religious leaders, cult leaders, bestselling authors of all kinds of junk theories, crank spiritual teachers, but also great philosophers, influential political leaders, prominent scientists, psychic healers, all sorts of researchers, etc. The System always produces thesis and counter thesis, heroes and villains, gods and devils, angels and fallen angels, saints and demons because competition is one of the main characteristics of our development.

Kundalini rising is a simplified way to visualize the transformation of the psychic, yogi, tai chi master, etc.

Kundalini rising is one of the modes of transformation.

The individuals undergo different transformation experiences, even different sequences of transformative events. Although there are some common features, each person undergoes the initiation and transformation process in a unique way.

The book *Kundalini Rising: exploring the energy of awakening is a compendium of articles on transformation* by various authors.

Stuart Perrin wrote something very important about Kundalini rising expectations.

"I meet many people who talk about the great kundalini experiences they had twenty or thirty years ago, but today they are often dried-out, aging, unhealthy people who live in the haze of memory instead of the creative and vital expression of the moment."

"When novices who don't have the proper education or guidance begin to naively and carelessly engage mystical experiences, they are playing with fire. Danger exists on the physical and psychological levels, as well as on the level of one's continued spiritual development. Whereas spiritual masters have been warning their disciples for thousands of years about the dangers of playing with mystical states, the contemporary spiritual scene is like a candy store where any casual spiritual "tourist" can sample the "goodies" that promise a variety of mystical highs."

In most cases "awakening kundalini" does not lead to transformation.

Spiritual awakening/enlightenment in general terms means entering into a world of doctrine, illusions, self-delusions, or beliefs empowered by the System, the great manipulator. Luckily, believing in Santa Claus not always is that bad.

Transformation does not always leads to "spiritual growing up." Some transformed individuals remain scientifically oriented; they are not spiritual, religious, or New Age proponents.

To become genius means to be transformed and work in cooperation with the System. Some people don't understand that they were transformed. They think that they had some physiological and psychotic problems (in many cases the symptoms are mild) which are actually part of the path to becoming a genius.

There is a little book titled *The Book: On the Taboo Against Knowing Who You Are* by Alan Watts.

Well, there is no taboo any more against knowing who we are: the conundrum is revealed—We Are Remotely Controlled Mortal Animals with early science, basic technology, and simple culture.

On the other hand, from a certain perspective, the Universe, life, the space civilizations, and humans are mere vibrating energy (all matter is vibrating energy) governed by information and high intelligence. Probably you want to be something more, but nobody on this planet can.

Chapter 19
Various Cases

In this chapter I present several cases suggesting existence of the System.

This case was presented in the book *Alien Dawn* by Colin Wilson.

Stanislav Grof is known for his early studies of LSD and its effects on the human mind. At a psychology conference (attended also by Colin Wilson) he presented the case of a young woman named Flora, a depressive with violent suicidal tendencies. She took part in a robbery in which the watchman was killed and she was sentenced to prison. After she was released, she became a drug addict and alcoholic. The young woman got in a series of troubles, including wounding her girlfriend with a gun while she was on heroin.

Grof, after a long hesitation, gave her medical treatment with LSD. During the third LSD session, suddenly her face froze into a mask of evil and she began to speak in a deep male voice. She seemed to undergo a total change of personality. The male voice presented himself as the Devil and ordered Grof to stay away from her because Flora belonged to him. Then the "Devil" shouted threats of what would happen to Grof, his colleagues, and the program if Grof continued treating her.

Colin Wilson wrote, "Grof says they could all feel the 'tangible presence of something alien' in the room. The threats showed an amazing insight into Grof's own personal life and those of his assistants. Flora herself could not possibly have acquired such detailed knowledge – she was not even a patient in the hospital."

Indeed, how would Flora know in detail the personal life of Stanislav Grof and his colleagues since she didn't know them? What was the mysterious source of information? Most probably she received it from the vast database of the System, which was playing the Devil. Some researchers might suggest that she was reading the minds of Grof and his colleagues, but telepathy does not occur directly between the minds of people. It goes through the System which controls it. Telepathy at will is impossible.

The case of Mary Schwartz

In her book *On Life After Death* the psychiatrist Elisabeth Kübler-Ross, a pioneer in the study of the near-death experience, tells the mysterious story of Mary Schwartz.

About 10 months after Schwartz's death, Dr. Ross decided to terminate the death and dying seminar. She was discussing this matter with the hospital chaplain Renford Gains as they approached an elevator. Then she noticed a woman who looked familiar. She was a bit transparent. Dr. Ross recognized her. It was Mary.

She approached Dr. Ross and asked her if she could accompany her to her office.

Dr. Ross wrote, "This was the longest walk of my life. I am a psychiatrist. I work with schizophrenic patients all the time, and I love them. When they would have visual hallucinations I would tell them, 'I know you see that Madonna on the wall, but I don't see it.' Now, I said to myself, 'Elisabeth, I know you see this woman, but that can't be.'"

Dr. Ross even touched her skin to see if it was cold or warm, or if the skin would disappear.

Dr. Ross, thinking she needed proof that this was not a hallucination, asked Mary Schwartz for a note with her autograph to pass on to Gains. In reality, it was a check of her sanity.

Mary Schwartz took a paper and wrote a note.

"Then she got up, ready to leave, repeating: 'Dr. Ross, you promise?' implying not to give up this work yet. I said, 'I promise.' And the moment I said, 'I promise,' she disappeared."

Dr. Ross checked the note. It was real.

The case of magic death

Douchan Gersi was a filmmaker, explorer, author, and adventurer. He spent most of his life in some of the most isolated regions on our planet documenting cultures he calls people of tradition. Gersi described his adventures in many books.

His parents were political refugees. They left Czechoslovakia, where he was born, and moved deep into the Belgian Congo (now Zaire), where he spent his childhood and part of his adolescence. Jungles and savannahs became his playground; tribesmen were his neighbors.

The next story is from his book *Faces in the Smoke: An Eyewitness Experience of Voodoo, Shamanism, Psychic Healing, and Other Amazing Human Powers*.

The watchman Moduku told the boy amazing stories about wildlife.

One day Moduku came to the boy and told him that the man who wounded his chest had settled nearby. He needed a black chicken for the ceremony of magic. Gersi agreed and delivered a chicken.

That night he made the journey to Moduku's small house. He looked wild as he crouched over the scattered feathers and bloody remains of the chicken. Gersi watched him make a small doll out of rags and grass. Then he covered it with thin feathers and blood. Moduku lit the contents of the calabash (hard shell of a fruit) and heavy smoke poured from it.

Slowly the smoke became very strange. Colors began to form bizarre faces—not human faces, nor those of animals or birds, but faces nonetheless.

Suddenly, he drove thorns into the doll: one in the head, one in the heart, one at the navel. Moduku had become so hysterical and frenzied, his body trembling all over, that the boy got scared and ran away.

The noises and voices of a gathering crowd awoke him early in the morning. He saw his father getting into the truck with a group of natives who looked frightened.

When his father returned he told that he had seen the watchman of the nearby sawmill dead, and yet still standing on his feet, leaning against the wall of his hut.

According to the father, only lightning could suddenly kill a man like that, leaving him in that standing position. But since there had been no thunder that night, the man must have been struck by a magic death.

When Gersi rushed to Moduku's small house he found it empty.

The case of Swedenborg

The widow of the Dutch ambassador in Stockholm, some time after the death of her husband, was called upon by a goldsmith to pay for a silver service which her husband had purchased from him. She was convinced her husband had been much too precise and orderly not to have paid this debt, yet she was unable to find the receipt. Because the amount was considerable, she requested Emanuel Swedenborg to ask the spirit of her dead husband about a receipt for the silver service. Three days afterwards Swedenborg visited the widow and told her that he had conversed with the spirit of her husband. The debt had been paid and the receipt was in a secret compartment in the bureau upstairs. When the secret compartment was opened, it contained private correspondence as well as the receipt.

The psychiatrist Stanislav Grof tells this story in *The Adventure of Self-Discovery.*

In one of his LSD sessions, Richard, one of Grof's patients, had a very unusual experience. It had an eerie luminescence and was filled with discarnate beings that were trying to communicate with him in a very urgent and demanding manner. He could not see or hear them; however, he sensed their almost tangible presence and was receiving telepathic messages from them. Grof wrote down one of these messages that was very specific and could be subjected to subsequent verification. It was a request for Richard to connect with a couple in the Moravian city of Kromeriz and let them know that their son Ladislav was doing all right and was well taken care of. The message included the couple's name, street address, and telephone number; all of these data were unknown to Grof and the patient. After some hesitation Grof finally went to the telephone, dialed the number in Kromeriz, and asked if he could speak with Ladislav. To his astonishment, the woman on the other side of the line started to cry. When she calmed down, she said with a broken voice: "Our son is not with us anymore; he passed away, we lost him three weeks ago."

The case of wonderman

Arnold Henskes, known by the pseudonym Mirin Dajo, was a Dutch performer. He adopted his stage name Mirin Dajo, which is based on the Esperanto for "wonder. " He let people pierce his body with "dagger-like objects." Dajo swallowed glass and razor blades.

In public performances at the Corso Theater in Zurich, he left audiences stunned—in plain view Dajo would have an assistant stick a fencing foil (rapier) completely through his body, clearly piercing vital organs but causing Dajo no harm or pain. When the foil was removed, Dajo did not bleed and only a faint red line marked the spot where the foil had entered and exited.

Dajo's performance was so nerve-racking that eventually one spectator suffered a heart attack, and he was legally banned from performing in public.

The Swiss doctor Hans Naegeli-Osjord learned of Dajo's alleged abilities and asked him if he would submit to scientific scrutiny. Dajo agreed, and in 1947, he entered a Zurich hospital. In addition to Dr. Naegeli-Osjord, Dr. Werner Brunner, the chief of surgery at the hospital, was also present, as were numerous other doctors, students, and journalists. Dajo bared his chest and concentrated, and then, in full view of the people present, he had his assistant plunge the foil through his body. As always, no blood flowed and Dajo remained completely at ease. By all rights, Dajo's vital organs should have been severely damaged, but he seemed in good health. Filled with disbelief, the doctors asked Dajo if he would submit to an X-ray. He agreed and without apparent effort accompanied them up the stairs to the X-ray room, the foil still through his abdomen. The X-ray was taken and the result was undeniable. Dajo was indeed impaled.

Finally, a full twenty minutes after he had been pierced, the foil was removed, leaving only two faint scars. Later, Dajo was tested by scientists in Basel, and even let the doctors themselves run him through with the foil.

According to Swiss physicians observing Dajo's performances and ultimately carrying out the autopsy on him (Brunner and Hardmeier), there were three aspects to be taken into consideration: painlessness, absence of infections, and absence of severe internal or external bleedings.

The case of the great Stockholm fire

Emanuel Swedenborg was a scientist, engineer, inventor, philosopher, theologian, mystic, and book author. At age 53, he entered into a spiritual phase in which he began to experience dreams and visions.

Here is a well-documented story about one of Swedenborg's visions.

On Thursday, July 19, 1759, a great fire broke out in Stockholm, Sweden. In the high and increasing wind it spread very fast, consuming about 300 houses.

When the fire broke out, Swedenborg was at a dinner with friends in Gothenburg, about 400 km from Stockholm.

About six o'clock Swedenborg left the company for a while and returned pale and in great agitation. Questioned, he said that a dangerous fire had just broken out in Stockholm and that it was rapidly spreading. He was restless and often went out into the garden. He said that the dwelling of a friend whom he named was already in ashes and that his own house was in danger of catching fire.

Two hours later, after Swedenborg had been out again, he exclaimed with relief, "Thank God! The fire is extinguished, the third door from my house!" Some of the guests were residents of Stockholm and had been greatly alarmed.

Word even reached the ears of the provincial governor, who summoned Swedenborg asking for a detailed recounting.

At that time, it took two to three days for news from Stockholm to reach Gothenburg by courier, so that was the shortest amount of time in which the news of the fire could have reached Gothenburg. The first messenger from Stockholm with news of the fire was from the Board of Trade, who arrived Monday evening. The second messenger was a royal courier, who arrived on Tuesday. Both of these reports confirmed every statement to the precise hour that Swedenborg first shared the information.

The case of the library angel

The following case is told by the medical doctor and psychiatrist Shafica Karagulla in her book *Breakthrough to Creativity: Your Higher Sense Perception.*

When Karagulla arrived in Edinburgh she called at the office of the medical superintendent of the Royal Edinburgh Hospital for Mental and Nervous Disorders. She didn't mention that she was researching psychic phenomena. She did not want to describe or discuss a book about the paranormal with anyone in the hospital. Years before in the library of the superintendent's office, she found a book with some information about energy fields around patients. She wanted to find the book again but she didn't remember the title and the author.

Karagulla told her psychic friend Kay about the book and both entered the library. She engaged the superintendent in conversation while Kay unobtrusively looked for the book, walking over to the bookcases and beginning to idly pass her fingers along the bookshelves. Kay didn't know the title nor the author of the book.

On the third wall Kay paused and pulled out a book. She leafed through it carelessly and handed it to Karagulla as she joined in the conversation with the superintendent. It was the right book.

Later at lunch when Kay and Karagulla were alone, she wanted to know how her psychic friend had found the book so quickly. Kay said that when she was looking for a book or magazine, she ran her fingers quickly along the bookshelf. When she felt a tingle on the end of her fingers, she had found what she was looking for.

Chapter 20
UFOs: Mind Control, Materializations, or Alien Spacecraft?

In 2020, the U.S. Congress ordered the Pentagon to produce a report on UFOs. The document was supposed to give detailed analysis of unidentified aerial phenomena data and intelligence that had been compiled by the Office of Naval Intelligence, the Unidentified Aerial Phenomena Task Force, and the FBI.

On June 25, 2021, the Pentagon released the much-anticipated unclassified report on UFOs, called "Preliminary Assessment: Unidentified Aerial Phenomena."

Officials examined 144 incidents from the past two decades, including videos showing UFOs.

Of these, 80 reports involved observation of UFOs with multiple sensors.

The videos confirm that in some cases the UFOs are material objects that can be video recorded.

The report has no explanations for 143 aerial phenomena incidents. Only one could be identified with "high confidence." That one turned out to be a large, deflating balloon.

Possible explanations included objects like birds, balloons, drones, atmospheric phenomena such as ice crystals or thermal fluctuations, new developments by US government or private entities, and technologies deployed by foreign states.

The document found no clear indications that there is any nonterrestrial explanation for the flying objects, but also did not rule it out.

The report does not contain the words "alien" or "extraterrestrial."

The document cited 18 incidents "that appear to demonstrate advanced technology" based on flight characteristics. In those incidents, the UFOs "appeared to remain stationary in winds aloft, move against the wind, maneuver abruptly, or move at considerable speed, without discernible means of propulsion."

Some of the objects were flying at seemingly impossible speed and doing seemingly impossible maneuvers.

In a small number of cases, military aircraft systems processed radio frequency energy associated with UFO sightings.

The document described the objects as a threat to flight safety and possibly national security.

The report was largely considered to be inconclusive.

It suggests further study or "pending scientific advances" may be needed to explain the UFOs.

In short, the militaries and their scientists said in the report: yes, there are some physical UFOs out there doing impossible things but we don't know what they are.

Luis Elizondo was the director of the now defunct Advanced Aerospace Threat Identification Program (2007-2012), a $22 million special access program initiated by the Defense Intelligence Agency in order to study unidentified aerial phenomena, also known as UFOs.

According to History Channel, when Luis Elizondo worked for the program, he outlined five observables, which he noted as commonly associated with UFOs:

1) Anti-gravity lift

This is the ability to fly without apparent means of propulsion or lift. In the USS Nimitz incident, the crafts were tubular, shaped like a Tic Tac candy.

2) Sudden and instantaneous acceleration

3) Hypersonic velocities without signatures

If an aircraft travels faster than the speed of sound, it typically leaves signatures like heat, sound, atmospheric ionization, electromagnetic radiation, sonic boom, and vapor trail. Many UFO accounts note the lack of such evidence.

4) Low observability, or cloaking

5) Trans-medium travel

This is the craft's ability to seamlessly move through space, air, and water.

Simply put, UFOs defy the known physical laws and present technology.

In the article "Tom DeLonge's Warped UFO Tour" published by Matt Farwell in *The New Republic* magazine, August 10, 2020, Elizondo told the author that there was also another "observable" they were more reluctant to discuss. "The sixth is biological effects."

According to Elizondo, UFO and paranormal experiencers were in danger of potential morphological changes to the body and brain.

"In one instance, a staff officer came home from a mission, went to their apartment, and fell asleep in the bedroom. The officer's roommate experienced what can only be described as poltergeist phenomena—mostly books flying off shelves—serious enough that the police were called."

Now, let's plunge into the sea of UFOs and alien lore!

According to the majority of science fiction scriveners, most planets are inhabited by nudist morons. In lots of novels and movies, the aliens are naked. According to numerous witnesses, many UFO extraterrestrials are naked, too.

The naked alien invasion folly started with the novel *The War of the Worlds* by H. G. Wells. Space invaders from Mars came to conquer the Earth…naked! The technologically superior tentacled astronauts bravely breathed terrestrial air and knew nothing about microbes. The nude Martian invaders died out because they couldn't resist the germs on Earth.

But can we resist non-terrestrial microorganisms, engineered germs, quasi-alive nanomachines, or ordinary alien microfauna and microflora that would normally live in the spacecraft and bodies of the extraterrestrial intelligent beings? The dead Martians of H. G. Wells' novel could cause a pandemic with fatal consequences for humanity. The terrestrial germs killed off the Martians, but the Martian germs in their bodies could exterminate all humans.

Wearing a spacesuit is not a matter of shame, moral standards, or fashion. It is to protect the fragile biological body from the hazards of the environment—harmful microorganisms, deadly gases, radiation, extreme temperatures, the vacuum of space, different atmosphere, sudden changes of pressure in the atmosphere, harmful liquids and vapors, fire, g-forces, some weapon attacks, etc.

How would the unprotected biological body react to the vacuum of outer space? Theory predicts and animal experiments confirm that after one or two minutes, the creature dies.

What would happen to a naked creature (vicious invader or wacky tourist) on a planet like Venus if it were not a native Venusian? It would die and be turned into ashes within a few seconds.

A battle spaceship has dramatic accelerations and decelerations, and without a special protective spaccsuit, the creatures would die or at least be unable to pilot the battleship or fight. Blood circulation becomes impaired because the heart must work much harder to pump blood through the body. The heavily impaired circulation leads to death or to the dimming or loss of consciousness, as the heart can no longer pump blood to the brain. Anti-g suits are a norm for surviving in a military aircraft.

The clothes are also a way for the pilots and crew to carry on themselves weapons, instruments, communication devices, medicine, food, personal belongings, etc., which are vital for survival in a hostile alien environment.

One might wonder from which cavity of the naked body an alien would produce a weapon, a screwdriver, or a cookie.

Spacesuits are covered with plenty of pockets and velcro, helping the astronauts keep everything they are working with nearby, for in weightlessness everything floats away.

Why do so many writers and moviemakers insist that aliens be naked?

The eternal conflict between good and evil, personified as God and the Devil, is one of the most conventional themes in the arts. God and the Devil are a mythic personification of reality with a huge impact on human society, so huge that we observe their supernatural manifestations in every aspect of man's life, destiny, arts, and history. People have been watching this classical tragedy in theaters and in real life since the dawn of civilized man. Writers from ancient times thematized the classical conflict in religious terms, and their fiction describes events as manifestations of the never-ending battle between God and Devil. Writers and audiences need these ultimate symbols of good and evil, God and the Devil, and their mighty invisible presence.

The evil naked aliens are the modern form of the ancient mythological image of the Devil. People want to hear and see the archaic tale about the go(o)d defeating the (d)evil over and over again. This makes them feel good.

People like to read mythological stories and watch movies with such elements. They are programed to do so.

Instead of independent, conscious thinking, we get mental Legos from our unconscious, which form the mental picture of the world, just like the colorful, interlocking, plastic Lego bricks can be assembled and connected in many ways to construct various objects. The unconscious mental Legos were called archetypes by Carl G. Jung. Of course, there are also other unconscious mental preforms, too.

All humans, without exception, are in the tight grip of their primitive unconscious, which delivers the mental picture of the world and serves the mundane, too. Humans are still very far from the moment of independent logical thinking. Instead, we get mental Legos mastered throughout the history of the

Universe, but the developmental pattern was created many universes back in other times.

Because of the evil nature of the Devil, writers and artists usually depict him as an abhorrent, repulsive creature. Stories, plays, paintings, etc., often show him naked with hooves or goat legs, various forms of horns, scaly or furry skin, fangs, and a tail. The cloven hoof has been associated with the Devil.

The Devil is often identified with the serpent that tempted Eve in the Garden of Eden. Adam and Eve were expelled from the eternal paradise of happiness where they were doing nothing but eating, gaping, and having sex. The tentacles (resembling a serpent) of the evil aliens are another symbol of the devil. Humans have a natural negative attitude about snakes. The Devil has a repulsive skin, as does the snake.

Medieval artists often depicted the Devil as a half-man, half-beast. Therefore, the evil aliens should also have beastly forms.

Fear of the Devil (evil) is imprinted in the unconscious of people. It is a powerful archetype in many societies.

The primitive, beastly, and uncultivated is naked. The sophisticated wear clothes. Some ancient gods are also naked, but the traditions have changed since antiquity and nudity is unacceptable for most people; it is a sign of primitiveness and shame, imprinted deep in the mind of humans. When we appear naked in front of other people in our dreams, we feel ashamed and embarrassed.

People are normal, aliens and the Devil are an aberration from the norm. People have hands, aliens have tentacles. The Devil can turn into a serpent, looking like a tentacle. People have legs, the Devil and aliens have hooves.

The Devil and aliens are generally regarded as the adversaries of God and are usually associated with danger, violence, and death, while man is created in the image of God.

The Devil and aliens are powerful destructive forces, God and man are creative.

But one should never forget that gods (the System and the superior intelligence) are always on the side of the faster developing civilizations.

Back in history, we see that the failed human competitors were deleted. Slow-developing groups and civilizations are deleted, too; if lucky, they are assimilated.

All species of the genus *Homo*, except for *Homo sapiens sapiens*, are extinct. The less developed predecessors, temporary peers, and interbreeders like *Homo neanderthalensis, Denisovans, Homo habilis, Homo erectus,* and all others were deleted by the System in order not to interbreed with humans and our predecessors.

Now and in the far future, it will be the same. Many civilizations of the Universe will be deleted.

Our bodies and our environment are teeming with microorganisms. It is normal to expect that the bodies of all living alien beings are also teeming with microbes, which are harmless for their hosts, but sometimes deadly for us. The *alienoses (exonoses)* are diseases communicable from extraterrestrial organisms (alien sentient beings, animals, plants, quasi-living beings, artificial biological organisms, or currently unknown forms of life) to humans under natural conditions. They will be a subject of (near) future medicine shortly after the first contact with alien life forms.

Normal flora are microbes living in and on the human body. There are more of "them" than "you" in you. Many billions of microbes live harmlessly on our skin and in the gut; we breathe them in and out. Numerous aerobic and anaerobic bacteria reside in certain human anatomical regions: in the lower intestine there are approximately 100 billion microorganisms per gram of fecal matter; in the mouth approximately 1 billion microorganisms per ml.

of saliva; in the nose approximately 20,000 microorganisms per ml. of nasal washing; on the human skin approximately 100,000 to 1 million microorganisms per cm^2, depending on the tested skin surface. After puberty, the vagina is colonized by *Lactobacillus aerophilus*. One or more of the herpes viruses infect nearly 100% of the adult population.

Over 400 distinct species of microorganisms inhabit the various regions of the human digestive tract, making up nearly 2 kg. (approximately four pounds) of every individual's total body weight.

Humans are symbiotic animals. This vast population of microorganisms far exceeds the number of tissue cells that make up the human body. We have about 10^{13} cells in our bodies and 10^{14} microbes.

The human body can't be fully sterilized because such a specimen would die shortly thereafter—some of these microorganisms take part in vital biological processes, while others keep our immune system fit.

The failure to detect alien microorganisms or alienoses (diseases transmitted from alien life forms to people) on Earth is a very strong argument against extraterrestrial visitors of biological origin.

Every year hundreds of thousands of UFO sightings, abductions of people by crews from other planets, medical examinations of humans in alien spaceships, sexual relations with creatures from other planets, various accidents, autopsies on extraterrestrials, and other E.T. contacts are reported.

There are many reports from all around the world describing men and women being taken aboard flying saucers and having sexual intercourse with various alien races, and even the birth of space babies. Microbes are transmitted by the direct transfer of bodily fluids, such as blood and blood products, semen, and other genital secretions from one person to another; they can enter the body through the lining of the vagina, penis, rectum, or mouth. The microbes can also be transmitted across the placenta.

Every day millions of samples of blood, human tissue, semen, saliva, urine, feces, etc., are collected and sent to labs to be analyzed. Not a single researcher, physician, or medical technician has ever reported an alien microorganism or alienosis.

There are persistent rumors that dead bodies of E.T. astronauts were found among the debris of the alleged UFO that crashed near Roswell.

Rescue, military, and medical personnel reported dead alien corpses at Roswell and other crash sites. Supposedly many extraterrestrial cadavers have undergone an autopsy examination. Witnesses claimed that in an underground base they saw a room full of canisters where bodies of dead aliens were stored. Graves of extraterrestrial beings have also been reported.

Since 1995, hundreds of TV stations all around the world have broadcast an alien autopsy film. Ray Santilli, a London-based film producer, claims to have bought it from a cameraman who took the footage in 1947 at the crash site near Roswell.

There are dozens of autopsy reports on alien cadavers recovered from various crash sites.

There are interviews with several medical doctors who performed autopsies on E.T. bodies.

The decomposing alien corpses at the crash sites, their blood, urine, feces, flesh, saliva, etc. are surely a source of extraterrestrial contamination. Healthy non-terrestrial astronauts are also dangerous. They could cause deadly alienoses. The autopsies and preservation of alien bodies after World War II were not safe enough by modern standards, and contamination with E.T. microorganisms was inevitable. In any manned alien spaceship there would be food, canisters with samples of tissues and microbes, medicine, drinks, breathing air, plants, hardware, all kinds of supplies, and so on, which are all sources of microbial contamination.

The governments on Earth could hide alien corpses, but they are not in the position to hide microorganisms left by extraterrestrial visitors. It's just impossible. No government (official, shadow, secret, mythical, mystical, or whatever), organization or individual on our planet has the technology to do that.

Where are all of the microbes left by the supposed aliens? It is highly improbable that extraterrestrial microorganisms are identical to those on Earth. Only one surviving bacterium or virus could multiply into billions in no time. There are trillions of various microbes in just one body (dead or alive).

According to numerous alien encounter reports, humans were in close contact with extraterrestrial astronauts and ufonauts, which are depicted as breathing terrestrial air, drinking water, eating human food, and so on, and most of them don't even wear adequate protective suits and helmets but only fancy skin-tight suits, silver suits, yellow ski suits, glowing aluminum suits, diving suits, jumpsuits, even Nazi military uniforms. Some aliens were dressed like humans or in some sort of mockery of space suits, but many of them were actually naked during the contact.

Millions of humans are reported to have entered alien spaceships, and none wore protective suit. Germ contamination is equally dangerous for humans and extraterrestrials.

Researchers still have not detected extraterrestrial microorganisms on Earth, and one can conclude that there haven't been any manned visitations (now or in the past) by extraterrestrial civilizations, or if there are any, they have been very limited in number and activities and are under strict control, having nothing to do with the UFO and alien abduction farce.

Humans, like all extraterrestrial civilizations, should create a reliable space-based surveillance and target acquisition system able to detect, quarantine, or destroy any alien life form that might be dangerous for humans and for terrestrial flora and fauna. The animals and plants on our planet are vital for survival of humanity.

The human immune system protects the body from pathogens, foreign substances, and malignant and infected cells by destroying them. We should create a similar protective system in the Solar System in order to survive.

Humans are still alive because the extraterrestrials, their alien microbes, their flora and their fauna aren't here on Earth.

For their own good, the space civilizations are separated by huge interstellar distances, but the phase of contacts and competition will inevitably come.

Space civilizations will encounter each other closely when most of them are in a position to survive such contacts.

In 1991, the Roper Organization polled about six thousand adults to find out how many thought they might have been abducted by alien creatures. According to the research, two percent of the American people, about 6.4 million Americans, believe that they were abducted by aliens. Extrapolating to the population of the world, worldwide about 160 million people were abducted by aliens? Hard to believe, because such an operation would require tens of thousands of aliens abducting people 24/7 and thousands of spaceships constantly landing on our planet and taking off, especially taking into account that most of the abductees were abducted many times.

A large number of people claimed that they were abducted by aliens and were medically examined by them. The abductees are convinced that these events are real because of the scars on their bodies that were the result of these examinations.

There are physiological, biological, and psychological effects on UFO witnesses. Some of the people, after contacts with aliens and UFOs, begin to experience personality deterioration. They have nightmares and peculiar hallucinations. UFO contact sometimes affects the sleeping pattern: in some cases witnesses sleep too much for about a week or months. UFO encounters can cause temporary blindness, numbness in fingers, nose-bleed, menstrual cycle disruption, red and swollen eyes for days, pain in the genitals, nervousness, scars, marks, and symbols on the skin (V-shaped, diamond-shaped marks on their bodies, red triangles, strange letters, etc.), nauseating waves of heat, swollen joints, skin infections, red welts on the neck and the body, witnesses sometimes smell a foul odor, lose consciousness, pains in the legs and arms, weight loss, vomiting, deteriorating vision, puncture wounds on the body that do not fade for months, sunburn-like blisters, nausea, diarrhea, discoloration of the skin, temporary paralysis (akinesia), a purple rash covers the body, and sunburns (they often appear in areas of skin covered with clothing). Some witnesses are unable to eat normally for several days; they get teary eyes, tingling sensations, electrical shocks, a feeling of heat, perception of odors, etc.

Many UFO witnesses and abductees are confused about time and ask themselves, "Why is time passing so slowly?" like in a dream when everything is moving in slow motion.

Abductions, stealing human babies, sex with nonhuman supernatural entities, and leaving scars on the bodies as a result of these events are as ancient as humanity. Modern researchers examine UFO witnesses and abductees for newly formed scars that should prove that these events are real, but priests and the Inquisition did the same to prove that the suspects practiced witchcraft or took part in Satanic rituals. Making scars on the human body has been used since the dawn of humanity to convince people that supernatural events are real. These mechanisms work today, too, and it should prove that UFOs and ufonauts are real extraterrestrial objects.

The eyes of many experiencers of UFOs and paranormal events were reddened and sometimes almost swollen shut. The splendid luminous UFO could be emitting some radiation which could harm the witness but not the mothman or some hairy large beast.

The System is using reddened, watery, and swollen eyes to convince us…in whatever it fancies!

It can make your eyes reddened, watery, and swollen when you see a terracotta garden dwarf. The System can make you converse with the terracotta dwarf who will reveal to you the greatest secrets of the Universe or of your friends and relatives, and will explain sacred spiritual teachings to you. It will know in great detail your past and future.

The System has total control over the human body, mind, and unconscious.

A regular stage magician can make you see and meet the entire menagerie of UFOs, close encounters of the third kind, alien abductions, deities and demons, mythological creatures, etc.

Some individuals have reported reddened eyes, strange scars, or wounds on their bodies, etc. after UFO events, which they consider a hard proof of their encounters with aliens. A stage magician can make the same marks on your body, too. He can put a coin on your hand and say that it is red hot, and you will receive burn blisters.

Man's subconscious is very powerful and it could cause various wounds, scars, illnesses, and even death.

Stigmata are body wounds that were not caused by an instrument or weapon.

Reported cases of stigmata take various forms: wounds of Jesus in the wrists and feet from nails; and in the side, from a lance; or wounds to the forehead similar to those caused by the Crown of Thorns. Individuals who have obtained the stigmata are described as ecstatics. At the time of receiving the stigmata, they are overwhelmed with emotions. No case of stigmata is known to have occurred before the thirteenth century.

Occult, mystic, and religious manifestations often include not only stigmata but also severe beating by invisible entities. People receive multiple bruises and sometimes broken hands and legs.

We can see that the human brain and mind are so constructed that they could be easily manipulated in order for individuals and humanity to be controlled by other humans and by some higher intelligence.

The human mind and body are so easily manipulated that it is as if there is an instrument panel attached to the human brain and body. Why are humans constructed this way? Most probably to be controlled.

If UFOs make impossible maneuvers like sharply changing direction at tremendous speed, defying the laws of inertia, or appearing out of nowhere and disappearing again, then reappearing again in another part of the sky, researchers assume that the alien civilization owns very advanced technology, making the impossible possible. But when witnesses see a UFO with a crew of vigorous human skeletons, or three humanoid alien visitors merge into one, or Nordic aliens in Nazi uniforms speaking in German, researchers ignore these facts. Flying saucer events quite often violate the known laws of physics and common sense. Actually, this is one of their basic characteristics. Just like the religious and paranormal miracles.

Many researchers try to avoid the word *hallucination* in regard to UFO encounters because it is connected with negative elements like pathology, drugs, alcohol, etc. And there is no alien spaceship but an illusion. Most UFO sightings are nothing but controlled hallucinations.

Hallucinations can be induced in perfectly healthy people, too. In the Fatima case about 100,000 people were seeing things that did not exist in reality, ergo, they were experiencing induced hallucinations. Humanity has a long tradition of controlled hallucinations. Since the dawn of humankind, people have been seeing deities, elves, angels, flying chariots, trolls, devils, flying witches, etc. The list is very long.

Even ordinary stage magicians can induce hallucinations.

Witnesses of UFOs, alien abductions, or mythological images state that before the phenomenon event, they often experienced a sense of dulling the consciousness, an odd sense of oppression, a dreamlike sensation, and a sense of dreamlike unreality. Their mind was prepared for embedding false memories or superimposition of images, sounds, thoughts, and emotions.

"Oz Factor" is a term invented by Jenny Randles in 1983 to describe the strange, seemingly altered state of consciousness commonly claimed by some witnesses of UFOs and other paranormal events. She noted the strange calmness and lack of panic described by the contactees relative to the bizarre circumstances that they described, and she said that they described and defined the Oz Factor as "the sensation of being isolated, or transported from the real world into a different environmental framework...where reality is but slightly different, [as in] the fairy tale land of Oz."

Many supposed UFO starships look more like huge toys with blinking multicolored lights intended to attract the attention of the witness and mesmerize them.

The UFO-encounter witnesses and abductees are in the realm of controlled visual, auditory, emotional, and thought hallucinations superimposed over the real picture of the world or false memories implanted in their mind.

The images and thoughts are superimposed on the real picture of the environment so that the individual can't understand what is real and what is additionally embedded. Superimposition is the placement of an image or video on top of an already existing image or video. The newly created memories in these cases are based on real events that are unfolding plus the superimposed ones. The UFO witnesses can't distinguish between reality and superimposed images, thoughts, and emotions.

Sometimes memories of nonexistent UFO or paranormal event is directly implanted in the mind of the witness.

Induced images and thoughts become part of the human mind and memory, and when a witness is hypnotized in order to reveal what he or she saw, they reproduce the induced pictures and thoughts that are already part of their reality.

In the movie *Total Recall* the Rekall company provides memory implants of vacations or other events. Quaid (Arnold Schwarzenegger) opts for a memory trip to Mars as a secret agent fantasy. During the procedure of memory implanting something goes wrong and the employees can't understand what was real and what was an implanted hallucination. Quaid starts revealing previously suppressed memories of actually being a secret agent. He is attacked in his apartment by Lori, who reveals that she was never his wife; their marriage was just a false memory implant.

The so-called psychological hypothesis claims that UFOs, alien abductions, religious, and mythological visions are a product of the unconscious. This notion became popular with the book *Flying Saucers: A Modern Myth of Things Seen in the Skies* by Carl G. Jung. This is only partially true because most of the images, emotions, and thoughts during UFO events do not origin in the unconscious but are created by the System and are embedded in the unconscious. Only part of the UFO and paranormal events are products of the overactive human imagination or mental pathology.

When Carl Jung coined and made popular the term *collective unconscious* he never explained the mechanism behind this phenomenon. Now we know that the System is behind the collective part of human unconscious. It is connected with the unconscious of all men.

Jung wrote, "My thesis then, is as follows: in addition to our immediate consciousness, which is of a thoroughly personal nature and which we believe to be the only empirical psyche (even if we tack on the personal unconscious as an appendix), there exists a second psychic system of a collective, universal, and impersonal nature which is identical in all individuals. This collective unconscious does not develop individually but is inherited. It consists of pre-existent forms, the archetypes, which can only become conscious secondarily and which give definite form to certain psychic contents."

Jung claimed that the collective unconscious is inherited, but how? The only biological inheritance system science knows is the DNA mechanism, but it can't transfer knowledge from parents to kids. Even if we suppose that information could be transferred from parents to kids, this is an individual exchange of knowledge and it cannot play the role of collective, common base of data for all humans on this planet. The archetypes and all other content that becomes available to the unconscious is produced by the System.

The great discovery of Carl Jung was that a large part of the unconscious, especially the symbolic one, is collective for all people on the planet. The new element to this theory is that the symbolic content is in a common reservoir that is connected to all individuals. At the center of this gigantic network is the System.

The controlled hallucinations are "artificial" and are not the result of pathology of the brain but of mind control. I have put the word artificial in quotes because this mind control system has existed since the dawn of humanity and became a natural part of the human self. On the other hand, the word artificial here is appropriate because humans still assume that their thinking is independent and uncontrolled.

The System is a natural part of the human mind and the self. It is the result not of the biological evolution of the humanity but of an external agency.

A large part of UFO sightings and encounters are implanted in memory as an integral entity or are superimposed over the reality. Such UFO witnesses successfully pass polygraph (lie detector) tests because they don't lie; they don't invent the extraterrestrial encounter or alien abduction events. The UFO events became an integral part of their memory and mind. In such cases, experienced polygraph operators usually say that the contactees and the abductees actually believe they did experience an encounter or abduction. But *believe* does not mean that these events really happened. The hypnotists in most cases assume that the "memories" are genuine. Many UFO witnesses and abductees agree to be hypnotized because they want to check the authenticity of their story.

It is extremely difficult to tell what is real and what is not when it comes to material retrieved through hypnosis. Only very experienced and intelligent hypnotists can reveal the true memories.

Some researchers use hypnosis to reveal the UFO and alien abduction events, expecting to get to the genuine nature of what really happened. But they should know that actually they are revealing the controlled unconscious of the person—memories are implanted, pictures, thoughts, and emotions are superimposed on real events. Even during the hypnosis sessions the mind of the hypnotist and the hypnotee could be controlled and manipulated.

Statistics reveal that about half of alien abduction cases were discovered under regressive hypnosis. The most impressive UFO encounters and abduction cases were reported by people under hypnosis. The alien abduction and UFO "memories" were inserted by the System into the unconscious mind of the "witnesses" and they were "revived" by the hypnotists. In some cases the abduction "memories" were actually created by the hypnotists from visual material memorized by the "abductee" from movies, TV shows, books, newspaper pictures and articles, etc. Some hypnotists actually plant suggestions into the mind of the supposed abductees and contactees.

Lucia dos Santos, one of the three young shepherds of the Fatima case, said that the day after the religious vision she had "no strength to do anything." Many UFO witnesses and abductees report the same lack of strength and confusion. They felt oddly drained and exhausted. Most probably this is a result of the manipulation of the conscious, the unconscious, and the subconscious.

Michael Talbot wrote in his book *The Holographic Universe*:

"Other UFO encounters are even more surreal or dreamlike in character, and in the literature, one can find cases in which the UFO entities sing absurd songs or throw strange objects (such as potatoes) at witnesses; cases that start as straightforward abductions aboard spacecraft but end up as hallucinogenic journey through a series of Dante-esque realities; and cases in which humanoid aliens shapeshift into birds, giant insects, and other phantasmagoric creatures."

People are designed and constructed in a way to be easily and effectively controlled and manipulated in order to develop as fast as possible.

What most impresses people taking LSD and other psychedelics is the sense of *reality* of what they are seeing. Most of the UFO cases also give a strong sense of *reality* of what the witnesses see. But in both cases these are mere hallucinations superimposed over reality or implanted memories.

Men In Black (MIB) are part of the UFO mythology. Often there are inconsistencies just like in the UFO and the ufonaut cases: some of the MIB hover above the ground or floor, they are gliding along the ground, they don't know how to use simple home things, they don't know how to eat certain foods, they wear brand-new clothing, they know too much or too little about humans, they walk through walls, etc. They are not part of some conspiracy of the governments on our planet but are a piece of the farce called UFO lore.

A sonic boom is the shock wave that is produced by a craft flying at a speed equal to or exceeding the speed of sound. It is heard on the ground as a sound like thunder. The supposed alien spaceships, despite their impossible maneuvers, accelerations, and speeds, never create sonic booms. Because they are not material objects in the sense that we understand material or are controlled hallucinations. Some researchers claim that they do not make a sonic boom because they have a different propulsion system. The sonic boom is made by any material object moving through the atmosphere when it breaks the sound barrier, not by the propulsion system. Meteorites don't have any propulsion system but they also create a sonic boom. Even if a washing machine is moving fast enough in the atmosphere, it will create a sonic boom, too.

And second, no living being would survive the tremendous accelerations and instant change of direction at an impossible speed or a sudden stop of the supposed alien spaceship. Even if we suppose that the aliens visiting Earth with their craft can somehow avoid making a sonic boom and don't get splattered over the instrument panel because of the tremendous accelerations and instant halts due to their being so technologically advanced, their moronic behavior and foolish conversations do not correspond to super-creatures able to perform such technological achievements and interstellar travel.

In his book *The Eerie Silence* Paul Davies writes about the moronic behavior of the putative aliens.

"Another giveaway was the banality of the aliens' putative agenda, which seemed to consist of grubbing around in fields and meadows, chasing cows or aircraft or cars like bored teenagers, and abducting humans for Nazi-style experiments. Not what one would expect of cosmic super minds."

Precisely a supermind is doing the antics. The System and the superior intelligence controlling our Universe, life, and all intelligence are superminds. They are stimulating our thinking, science, technology, etc. We have to understand and prove what is hiding behind the UFO and alien abduction games.

Some researchers proudly write that a particular UFO event was observed by five, sixty, or a hundred witnesses, considering that the numerous witnesses are proof that the UFO sightings and the aliens are genuine. This is a common mistake because the Fatima phenomenon was observed by 100,000 witnesses, but this does not prove that the sun was dancing in the skies and changing its colors, nor that it was falling and was about to crash into the Earth. When it comes to phenomena, the large number of witnesses is not proof that a given event is genuine.

Many researchers, including the most famous ones, often stress the fact that the witnesses of UFOs, alien abductions, religious miracles, encounters with mythological creatures, etc., are very sincere, and that gives a great weight, value, and credit to their stories. But when it comes to mind manipulation, sincerity is just part of the trick. These people are not inventing their stories and they have no reason to be insincere. Contactees are convinced that they saw UFOs and they are sincere, and they really have in their mind visual, auditory, thought, and emotional input. It's true that they saw something, but they saw it with their inner eyes or under hypnosis, and there were no aliens and spacecraft in reality. Sometimes the UFOs are materializations but they are not spacecraft from distant planets.

Sincerity is not proof of genuine contact with extraterrestrial people or the Virgin Mary.

There is no such thing as reliable witness when it comes to UFOs, alien encounters, and other paranormal phenomena. People can be manipulated easily.

Even skeptics were very impressed with George Adamski's apparent sincerity. He was the first, and most famous, of several so-called UFO contactees who came into prominence during the 1950s. He described in great detail his contacts with Venusians in his books *Flying Saucers Have Landed*, *Inside the Flying Saucers*, and *Flying Saucer Farewell*. The aliens had taken him to their home planet Venus, and after that they visited the Moon and Mars. Adamski wrote that there were lakes, forests, snow-covered mountains, and lunar cities on the Moon where the local citizens strolled down the sidewalks.

On September 14, 1959, the Soviet probe *Luna 2* successfully landed on the Moon. It was the first man-made object to land on another celestial body. The same year, *Luna 3* orbited the

Moon and sent back pictures. There were no Lunar cities and Lunarians, no lakes and forests, or breathable air. Only barren rocks, craters, and dust. Adamski accused the Russians of falsifying the Moon pictures to deceive the Americans.

What about his visits to Venus? Many other contactees claim that they also visited Venus on board flying saucers and they saw there a very advanced civilization. Scores of aliens said that they are Venusians.

The Venusian atmospheric surface pressure is very high, 93 bar, and the human or alien biological body would shrivel almost instantly, not having time to take even a single breath of the deadly, blazing air of carbon dioxide and sulfur dioxide. While shriveling, the biological remnants of the visitor would catch fire, for the temperature is 462°C (863.6 degrees Fahrenheit). Then the ashes would begin to dissolve in the sulfuric acid. And this is the end of the Venusian trip!

The Adamski example is a good lesson to all researchers: never fully trust witnesses' testimony, no matter how sincere they are. Sincerity is no proof that what witnesses "saw" is real. It is part of the games of the System which controls the minds of people and produces material-like alien spaceships.

In some cases solid-looking or metallic-looking UFOs don't appear on the radar screen.

Some researchers think that observing UFOs on a radar screen is solid proof for the existence of these objects. But seeing UFOs on the radar screen is the same as seeing them in the backyard. Seeing UFOs on the radar screen is no proof that there are UFOs in the skies. Of course, there are genuine UFOs, but they are not extraterrestrial spaceships, and some of them could be watched on the radar monitor and even recorded. These UFOs are materializations that imitate alien spacecraft. The System can control not only the minds of people but also our material world.

There was a big UFO wave in 1968-69 in Vietnam, which included an epidemic of phantom airplanes and helicopters. On several occasions, military forces on both sides fired at the objects without effect. Why did the artillery, rocket, and machine gun fire have no effect? Because the aircraft were controlled hallucinations or materializations that only looked like real airplanes or helicopters. You can see such objects, including on radar, but you can't shoot them down.

In many cases you don't know: is this a hallucination or a materialization?

What leaves the burned circles and deep impressions found at some supposed sites of landings? In some cases, they are made by witnesses who don't realize what they are doing; in other cases, they are superimposed images. The Brazilian Paulo Gaetano turned off his car engine, opened the door, and got out of the vehicle but he told his companion that a flying saucer cut off the engine, a red beam of light projected at the car caused the door to open, and he was abducted by aliens. Witnesses can do a lot of things that support UFO contact or alien abduction, but they would not remember or understand what they really have done.

In many cases, the landing marks are hoaxes.

Of course, the System can also control the matter and some of the phenomena are material, but in most cases we have mind control.

There is no need for all UFOs to be material, imitating alien spacecraft. It is enough for people to believe that they are extraterrestrial spaceships. One just has to control the minds of the people and the mass media.

Sometimes UFOs were observed on radar monitors and even recorded (they are materialized objects), sometimes they are observed by witnesses but were not seen on radar—here we have several possibilities: they were material but invisible on radar, they were controlled hallucinations, they were real extraterrestrial spaceships using stealth technology (as some UFO buffs would suggest).

The System can control matter and sometimes it materializes various objects.

Lyall Watson was a South African botanist, zoologist, anthropologist, and author of many books. The following case of materialization and poltergeist activity is from his book *Beyond Supernature*.

In 1974, Watson was in Indonesia to investigate the strange events in the home of some salt-maker.

Watson and the family sat around a rough wooden table. The boy suddenly screamed. His right hand began to bleed from fresh punctures.

"The [kerosene] lamp flame turned blue and flared up, and in the suddenly brighter light I watched a cascade of salt pour down over the food. It wasn't a sudden deluge, but a slow and deliberate action which lasted long enough for me to look up and see that it seemed to begin in mid air.

"There was a slow cracking sound, as though the thick wood was tearing itself apart; silence, then a series of sharp raps, the sound of urgent knocking at a door; then it began to wobble. The family got up in a hurry and we all watched in horror as the heavy table heaved and bucked like the lid on a box containing some wild animal, and finally flipped over on its side. I joined the others and ran."

They stood outside for a while and when nothing further seemed to be happening, the they went back in. Watson had by now recovered some presence of mind and enough scientific curiosity to look very carefully at the table, walls and ceiling for any signs of strings, pulleys and other devices. He found none. Watson was certain by then that he was watching a typical violent poltergeist in action.

During his stay, Watson witnessed stones flying around inside the house in all directions. One, about the size of a walnut, struck him on the chest.

Michael Talbot reported his own experience with materializations in his book *The Holographic Universe.*

The materializations started when he was six years old. Inexplicable showers of gravel rained down on the roof at night. Later it took to pelting him inside the home with small polished stones and pieces of broken glass with worn edges. Sometimes various objects materialized like coins, a necklace, and several odder trifles. He usually did not see the actual materializations, but only witnessed their aftermath.

"On a few occasions, however, I did see objects actually materialize. For example, in 1976 I was working in my study when I happened to look up and see a small brown object appear suddenly in midair just a few inches below the ceiling. As soon as it popped into existence it zoomed down at a sharp angle and landed at my feet. When I picked it up I saw that it was a piece of brown drift glass that originally might have been used in making beer bottles."

Some of the UFOs, aliens, some paranormal entities, and ordinary objects such as salt, coins, pebbles, etc. are materializations of the System.

On June 29, 1954, Stratocruiser Centaurus of the British Overseas Airways Corporation left New York's Idlewild Airport bound for London with 51 passengers aboard. Four hours later at sunset, just after 9:00 P.M. the aircraft passed

170 miles southwest of Goose Bay, Labrador, flying northeast at nineteen thousand feet. The crew suddenly noticed an object to the west of the plane that was roughly five degrees aft of the port wing at a distance of about five miles. As they drew closer, the crew could see a large, pear-shaped UFO flying in formation with six other smaller objects.

Against the sunset these objects appeared dark, with the six smaller UFOs changing their positions around the larger craft which seemed to change from its original pear shape to a telephone handset shape and, in the words of Captain James Howard, a highly respected former RAF Squadron Leader with 7500 hours commercial flying on 256 Atlantic crossings to his credit at the time of the sighting: "What looked like a flying arrow—an enormous delta-winged plane turning in to close with us." The anomalous, shape-shifting object seemed to be solid with definite, clearly defined edges.

As the jetliner drew closer, the smaller UFOs formed into a regular line and seemed to merge into the larger object. Then the remaining large UFO appeared to suddenly shrink and vanish right before the eyes of the startled crew.

Captain Howard remarked later: "…they were obviously not aircraft as we know them.

All appeared black and I will swear they were solid."

"There is no question that this was not an illusion...and that it was being intelligently handled."

The crew and some passengers watched the UFOs for about 18 minutes.

There are many cases in which UFOs and aliens shrink and vanish on the spot like a TV image when the set is turned off or they slowly fade away.

The materialized UFO can produce various traces. In his book *Physical traces associated with UFO sightings* Ted Phillips made a catalog of 561 reports of cases when there were all sorts of traces after UFO events, but still nothing alien was found.

In 1954, a gyrating metal disk appeared over the city of Campinas, Brazil. Hundreds of witnesses reported that it dribbled a stream of "silvery liquid" into the streets. Government scientists collected some of this stuff, and Dr. Risvaldo Maffei later announced that it was tin. Again, nothing alien. Analysis of material left after UFO events showed to be composed of calcium, aluminum, ferrochromium, silicon, iron, zinc, tin, and other mundane elements. UFO researchers even got to investigate extraterrestrial pancakes. Again, nothing alien.

Even hallucinatory UFO events can leave traces which are actually staged by the System.

The System can dematerialize every object and creature that is in this Universe. Humans are no exception. It can also transmogrify and transform every object and creature. Or burn them to ashes. Or levitate them. Or translocate/teleport them. Everything is possible in a virtual world or in a material Universe driven by a simulation. If one has control over the simulation, he can travel from one end of the Universe to other instantaneously, defying the natural laws which are just computer/program rules. For now only the System and its master have control over the simulation and the material world.

In such a world you can't be sure of anything. It's difficult to say what is real, what is a controlled hallucination, and what is manipulation of space-time.

Some people believe that the US government is in contact with aliens and there is an agreement in which the extraterrestrials were given military bases (in most cases they mention Area 51) and permission to mutilate cattle but also to abduct humans and examine them medically, in exchange for teaching American experts about alien science and technology. An additional source of alien technology are the crashed extraterrestrial spacecraft, like the one at Roswell in 1947.

But we can't see any advanced alien technology in the contemporary USA. American technology, including that of military, was supposed to be many generations ahead of current science and technology because of their advanced alien teachers and retrieved extraterrestrial technology, but it is not very different than that designed in Russia, Europe, India, or China. For years, NASA even had to buy seats to the International Space Station from Russia. Current rocket technology, including the Moon landings in the past, is based on German technology from WWII. The Mars visitations now are using the same ancient technology.

Scientists from all countries are learning from each other and, of course, are stealing each other's secret technologies and discoveries. Modern science and technology are developing gradually. Despite the claims of Colonel Philip J. Corso and some researchers that lasers, transistors, microchips, fiber optics, etc. were retrieved alien technology, the educated scholar can follow in scientific publications their slow and painful advance, which continues today, too. The new technologies are progressing step by step, and there was no quantum leap, supposing the use of super-advanced alien knowledge.

And there is still no interstellar spacecraft technology on Earth. Our piloted spaceships still can't reach even the nearest planets in our Solar System. They will soon reach the planet Mars, but they will be using WWII technology. Now the USA, Russia, China, and India are constructing primitive heavy chemical rockets to reach Mars. These vehicles have nothing to do with the supposed super-advanced alien spaceships visiting Earth. An agreement between some chosen earthlings or governments and sophisticated aliens is just one of the many modern myths. And no state-of-the-art alien technology was retrieved from crashed flying saucers. And there are still no robots produced on Earth that can communicate with humans telepathically as the alien robots from UFOs can. And we still do not use muonic scalpels while levitating for bloodlessly removing cow's rectums, genitals, tongues, eyes, and ears as the aliens do so often.

The System creates religious and UFO events using the same elements. In many cases the UFO, the religious, and the paranormal are mixed, to the great disappointment of UFO buffs who believe that UFOs are alien spacecraft. UFO sightings and alien contacts are intended to be more technological and space oriented than religious and mythological events in order to stimulate scientific analysis and research; they are not religion oriented, and one should hardly expect the appearance of new faith centered on the UFO belief. The two phenomena have different goals and different development no matter that they use similar elements for manipulation of people's minds. Religion is strictly faith oriented.

Here is short list of such elements: reading the minds of the witnesses of the events; prediction of future events; rotating disks with many lights on them; the clothes of the witnesses, the ground, and trees become dry despite heavy rain minutes before; witnesses receive all sorts of "important" and in many cases absurd messages concerning the future of humanity and its demise or survival; body paralysis; various images, thoughts, and emotions are projected into the minds of the witnesses; objects, humans, aliens, and entities can appear out of the blue and disappear instantly; light beams and pillars of light; loss of sense of time; miraculous healings; symbols, scars, wounds, and marks appear on the body; telepathic communication with ghosts, aliens, religious figures, and extraterrestrial robots; loss of speech; uncontrolled singing and dancing of groups of people; contactees visit alien planets or strange other worlds, sometimes psychedelic; taking part in occult rituals, balls, and initiations; understanding and speaking languages (including extraterrestrial) that the witness never learned; falling of angel hair and rose petals; witnesses find themselves in some sort of strange cloud; levitation; automatic writing; walking through walls; smelling incredibly pleasant or abhorrent odors; transforming of entities, for instance, an ancient male god turns around and becomes a beautiful goddess, an alien turns into a bird; poltergeist events; merging of several extraterrestrials, deities, or UFOs; missing time; sudden stillness; development by some of psychic abilities after an encounter; they know everything about the witnesses, their relatives, and all human history; glowing eyes, and many more.

The glowing eyes are a sure sign that you are seeing a hallucination or materialization, not a real creature or extraterrestrial because such a critter is actually blind. Normally functioning eyes detect light and convert it into electro-chemical impulses in neurons. They do not emit light. Glowing eyes are one of the many elements the System uses to frighten humans with paranormal tricks.

Some authors and scholars proposed that our ancestors were genetically manipulated by ancient astronauts in order to accelerate evolution. Some estimates suggest that about 10,000 BC the world population was roughly between one and ten million. Can you imagine a planetary hunt by aliens catching one by one all humans on Earth in order to manipulate them genetically? It is impossible for an extraterrestrial crew of one or several spaceships to catch and genetically manipulate all prehistoric men on our planet. Where are the unmanipulated people? There should now be two kinds of people on our planet: genetically manipulated (enhanced) people and "wild" people.

Only the System can control and genetically manipulate all people, animals, and plants on Earth, and people will not even know about such manipulations.

A viable UFO theory should explain all phenomena like chariots of the gods, ghost rockets, airships, and UFOs, and the accompanying events like paranormal healings, levitation, prediction of future events, full knowledge of all human affairs and languages, details of the lives of all individuals, etc. The extraterrestrial hypothesis cannot solve the riddle.

The phenomena reflect the time they were active: in ancient times there were chariots of gods, in the 19th century there were airships made by ingenious inventors, in our time there are extraterrestrial spaceships with aliens from deep space.

UFO buffs have been collecting stories for many years, concentrating on the kinds of data that fit their theories. A good scientific theory about UFOs and the paranormal should be able to explain all collected data of the subject. Now most UFO researchers are rejecting data that don't fit their theories.

They reject the data of the moronic behavior of most of the alleged extraterrestrials. The UFO aliens' actions do not look representative of a supercivilization capable of interstellar travel. They reject the instances of poltergeist events after UFO encounters. They hate cases of aliens and people walking through walls and closed doors. They reject the cases of witnesses becoming psychic after UFO events. They hate fulfilled predictions of future events often associated with UFOs because this contradicts the contemporary science paradigm. They reject many other data that don't fit their theories.

The UFO and ufonauts phenomenon are only one fragment of much larger phenomenon fabricated by the System.

What are the UFOs and ufonauts? Part of them are controlled hallucinations or memory imprints done by the System.

In some cases the UFOs and ufonauts were created during sessions by hypnotists from visual material from movies, TV, books, newspaper pictures and articles, etc. Some hypnotists actually plant suggestions into the mind of the abductees and contactees.

Some UFO and alien visitors events are the result of mental pathology.

There are also many cases of overactive imagination.

Some UFOs and ufonauts are physical objects but not extraterrestrial spacecraft. They are materializations of the System.

Of course, we should not exclude the possibility that there are some technical devices resembling spaceships that travel across deep space and visit planets with goals unknown to us. They could be products of the superior intelligence owning the materialized simulation we consider "our" Universe. We should, of course, not exclude the possibility that there are some real extraterrestrial visitations from our space brothers, but they should be very limited in number and activities, and they have nothing in common with the UFO frenzy.

There are numerous reports that some strange radiation emitted from UFOs is stopping car engines. Experts explain this happened because of the radiation of the propulsion system of the flying saucer.

Car engines stopped by the radiation of flying saucers is an improbable explanation, for there should be thousands of UFOs engaged in cutting off the engines of cars all around the world. Here we should add the enormous quantity of aliens and extraterrestrial spaceships that are engaged in abductions, genetic manipulations, medical examination of millions of humans (including anal and vaginal probing, which are favored among aliens), rapes, removing and implanting of fetuses, sexual intercourse with humans, implanting tracking devices, removal of eggs or sperm, cattle mutilation, chasing people's cars, etc.

To fulfill this enormous alien operation there should be constantly on Earth tens of thousands of space visitors and thousands of extraterrestrial spacecraft.

Since WWII there have been tens of millions of UFO encounters and alien abduction events worldwide.

Why don't the UFOs cut off the engines of airplanes? There are a great many reports of UFOs flying near all sorts of aircraft. There is little difference between most plane engines and car engines. Why does the hypothetical radiation emitted by the flying saucers not stop airplane engines? Probably because it would the impossible to explain how the engine of an airplane stopped for ten minutes or half an hour but the aircraft stayed aloft all the time.

Witnesses also report that their car engines stopped when they saw big hairy humanoids (Bigfoot). Their manifestations are associated with UFOs because they were dimming the headlights, making radio static, and stalling cars.

But the large hairy beasts don't have a propulsion system, they don't emit radiation, and they don't have powerful magnetic fields, causing radio static, dimming headlights, and stalling motor vehicles. Why did the cars stop? The System produces paranormal manifestations (including the unexplainable stalling of motor vehicles) in order for the mysterious events to be witnessed and remembered, and for people to believe in UFOs, aliens, bizarre entities like Bigfoot, angels, etc.

After observing paranormal phenomena, researchers have to analyze them and create an adequate picture of our world.

The next big move probably will be the future posthumans, aliens, smart machines, and AIs saving themselves (leaving our controlled Universe), if it is possible.

It is also probable that in the far future, most advanced space intelligences (including posthumans, if they survive) will merge with the System, becoming a single mind, or they will become part of a hive of super-intelligences. The far future of humanity and other space civilizations is unknown to us. We hit a singularity wall which states that contemporary people can't know the distant future.

In *Proceedings of the First International UFO Conference*, 1980, Curtis G. Fuller compiled and edited articles from various UFO researchers. J. Allen Hynek wrote the article "What I Really Believe about UFOs" in which he stated:

"Let's not fall into a similar trap, trying to explain everything by our present knowledge. Every generation thinks that those who have gone before were sort of dumb. We forget that we're going to be the "ancients" to people who live in the fortieth and fiftieth centuries. They're going to look back on us and say, "They were really dumb in those days. They didn't even know what UFOs were.""

Indeed, we don't know the physics behind the UFO and ufonauts phenomenon, but we know what produces and controls them—the System.

UFOs are not alien spaceships visiting Earth to explore our planet, forms of life, and our civilization. The ufonauts know in detail everything about us, including our names, relationships to each other, our health problems, our modern and dead languages, our little and big secrets, our past and future, everything. They are here to teach us.

The UFOs and the alien visitors are here not to study us but we to study them. As a result we should create a realistic picture of our world.

Afterword:
designer civilization in a designer world

Spirituality is only in the minds of the people living in our technological Universe.

There are numerous books giving advice on how to advance to a higher level of spirituality, giving us numerous examples of spiritual feats. What is it to be a real master of spirituality? What can someone hypothetically do once they reach the highest level of spirituality? The supposed master of spirituality possesses the power of mind over matter and over the minds of other people. He could rearrange the furniture in the house with thought only. He could levitate over beautiful landscapes for fun. There will be no need to travel long, tiresome hours to visit beautiful landscapes or to other desired places because he can translocate instantly. Masters of spirituality will know the future, and most importantly, they can change the future, making their lives an endless super bliss. They will communicate telepathically with all critters (humans, aliens, animals) and plants in our Universe and far beyond. They could travel in the past and visit all interesting events, battles, or persons from previous times. They will see with their third eye everything on this planet, even throughout the Universe. For them there will be no need to eat (and produce foods) because they will consume prana—the endless free cosmic energy. If they get some disease—no problem, they'll just visualize that they are healthy and the disease disappears. The masters of spirituality will live forever in a paradise. They can know everything without learning because they are connected to the Source, the Cosmic Consciousness. On the other hand, they do not need to know anything because they are happy and are in paradise. No diseases, forever young. Man will be immortal and beautiful. All will be like magicians. There will be no need to work.

Yes, many such feats are possible and the System demonstrated that many times; they are not the result of some spiritual mastery but of the superior technology of the System. Everything that happens in our Universe is a result of sophisticated technology created by a superior intelligence. The future of humanity is technological, too.

In the distant future most of these "spiritual" feats will be possible and everyone will enjoy them. People, however, will be not highly spiritual but highly technological.

Today's magic is tomorrow's technology.

Researchers trying to create a model of our world and explain the phenomena forget that our world is becoming technological. The future is technological, too, and we should think in scientific and technological terms—not religious, folkloristic, mystical, mythic, New Age, spiritual, etc.

Our future bodies and minds will be technological products, too. The universe, animals, and intelligent creatures are evolving and becoming more complex and more sophisticated. Until now, the System was shaping humans. In the times to come humans and other intelligences will be shaping the Universe, their bodies, the animals, and the plants. The sophisticated technology will rule the world. Even the miracles we experience now are a result of the technology of the System. The future technology will be so advanced and complex that we can't even imagine what it could be. It will be very, very different from primitive contemporary technology. It will be intelligent, thinking. It will contain incredibly sophisticated virtual worlds and intelligences. At some point the future creatures will be able to create universes, animals, and people like us.

And these humans will wonder who created them, why their world is so far from Eden, what are the paranormal phenomena and the UFOs, why were such primitive creatures like them created, etc. They will live in a world which they will not like (just like us) because it will be designed to stimulate their fast evolution, not to please them. The same way the System is pushing us to develop as fast as possible. They and we are just insignificant short-lived critters, just fleeting phases of the developing intelligence.

Paradise, perfect happiness, infinite life, resurrection from the dead of our loved ones or of famous historical figures, eternal youth, and other wonders of this kind are perfectly possible even today. The model of a desired world could be materialized by the master intelligence. We could live in supreme bliss forever in Eden, but our world is guided by another model which requires growth, evolution, progress, a huge number of hard-competing individuals, groups of people, civilizations, and countless births and deaths.

Mortal, short-lived animals like humans cannot understand what immortal is, what eternal is, what infinite is! We should use such words very cautiously. It's impossible to truly understand them. We can't comprehend an immortal intelligence.

Our patron civilization is not going to save us from the benefits of the competition; benefits from its point of view that are only disasters for us. It does not intend to prolong our lives (we should take care of this ourselves) or to solve the poverty problem (another mighty stimulus), to stop the wars and crime, and so on. The story of the Savior that all are awaiting is just a myth, giving us hope for better days. It is counterproductive, for if it becomes a reality, it would slow down evolution. But the Final Judgment is a reality that humanity will inevitably face. Not all space civilizations will make it in the future.

Our patron civilization is furtively guiding us in a highly clandestine manner, revealing its presence to us through mythology, religions, parapsychological manifestations, UFO lore, various phenomena, etc. The goal is numerous intelligent species, developing as fast as possible.

In order for humans to evolve faster, the superior civilization creates a rich and often contradictory intellectual environment that should stimulate man's thinking. In the past, educational toys were in the form of gods, devils, angels, fairies, demons, trolls, etc. In the 19th century the airships observed in America were added, after that in the skies appeared airplanes with impossible characteristics, while in recent times we were invaded by ufonauts with their flying saucers, often abducting earthlings. In the past thousands years there were many cases of phantom horsemen who appear out of nowhere and then disappear into nowhere. Nowadays you can chase phantom automobile, men in black, or aliens that will disappear into thin air. All that is part of the educational environment of man.

It is also possible that humanity is just an educational toy of some obscure entity or of the evolving System.

The Drake equation is a probabilistic argument used to estimate the number of active, communicative extraterrestrial civilizations. It is fundamentally wrong because it suggests that everything in our world is a result of chance and physical laws.

Accepting the existence of superior intelligence and its tremendous influence on the Universe, life, and intelligence is the only way to create a reliable scientific picture of our world.

Maybe our patron superior intelligence is guided as well by a higher agency, and there might be some sort of multilevel creation and control.

In order to understand our Universe, life, and civilizations, we should comprehend the philosophy, ideas, strategy, and the works of the System and the controlling superior intelligence—a challenging task for the humble human mind.

Some researchers are writing about the great silence of space suggesting that humanity is alone.

Wrong, the superior intelligence never stopped communicating with people: through mediums, psychics, alleged aliens, saints and gods, and so on; it is producing UFOs, paranormal events, religious miracles, etc. People just don't understand these phenomena believing in ghosts, communication with dead people, all sort of demons, gods, angels, extraterrestrials, etc.

Now it's official: we are not alone! There is a superior intelligence that created the Universe, the System, and us. And it is constantly contacting us.

Listen! It is talking to us in various disguises!

Armed with the concept that we are living in a material world driven by a simulation controlled by a sophisticated data processing System which was created and is guided by a superior intelligence, you will now be able to explain by yourself the paranormal phenomena, the biological and social evolution of humans, UFO events, etc. This is the most fundamental and powerful idea explaining the existence and evolution of the Universe, life, and intelligence. Armed with this new concept, you will see the world with new eyes.

The secret of the philosopher's stone is revealed, too: this is the power to control the simulation. If you can control the simulation, you can do and have everything—you can turn lead into gold, water into vine; you can enjoy eternal life in perfect health; you will know everything; you can have the most attractive beauties of the past, present, and future. You can defy the laws of physics (which are only algorithms of the simulation) and travel great distances across the Universe instantaneously. No wonder that the UFOs are defying the physical laws—someone can control the simulation, hence our material world.

With sufficiently advanced technology, some individuals could take control over the simulation, reaching the long-sought goal of the alchemists for obtaining the philosopher's stone, which is control over the simulation. With the fast advancing science and technologies, the search for the philosopher's stone continues with renewed vigor.

Now we are closer to a more realistic picture of our world and how it works. On the other hand, the real picture is still well beyond our comprehension.

"All our science, measured against reality, is primitive and childlike." Albert Einstein

Printed in Great Britain
by Amazon